FAST FOOD INGREDIENTS REVEALED

What <u>Are</u> You Eating?

By

Mel Weinstein

DISCLAIMER

The purpose of this book is to provide readers with an educational guide to the typical ingredients found in American fast foods and is not intended as an exposé or critical analysis of the fast-food industry or the specific restaurants mentioned in the book.

"They know not what they eat."

--- Hieronymus Anonymous (fictitious philosopher)

PREFACE

In the course of human history, restaurants are a relatively new phenomenon. In the roughly 6000 years of recorded history, the first mention of a modern-day restaurant occurred in 1835 in Paris, France (at least in writings of western authors). The word "restaurant" originally referred to a broth with medicinal and digestive benefits. Places that sold healing broths were called "health houses." Of course, today it's rare to associate restaurants with healthy eating establishments, unless they are marketed as such.

So, it was only about 186 years ago that modern restaurants arose in western society. What about the other 5814 years of recorded history? How did people eat? Very simply, they gathered food and prepared it at home. If they traveled, they either carried food with them or foraged for it. In the more advanced societies, travelers might stay at inns or taverns, but the food offerings were slim and meager. The travelers had only a single choice: whatever the inn was offering that day. This means that, generally speaking, people of earlier times knew exactly what was in their food because they personally obtained it and prepared it. Contrast that situation with today's modern societies, such as the United States. Probably, these days, more people eat out than prepare meals at home. Even when eating in, people are more likely to prepare commercial, packaged foods than whole, fresh foods. As a consequence, most people don't know what they're eating, and, even worse, don't care. This ignorance gets even worse when people eat out. Restaurants, either local, regional, or national, are under no obligation to reveal the ingredients in their food offerings. Some do, out of a social obligation, but the vast majority don't. When it comes to eating out, we are very trusting souls. We trust that eating establishments are not feeding us contaminated foods or foods containing harmful ingredients.

However, with the rise of the modern food industry and the clever work of food scientists, restaurant food has radically changed since 1835. When you go to a fast-food restaurant, do you know what you're eating? Not likely. The reasons are: (1) You only care about availability, cost, and taste or (2) The information is either not available or difficult to find. The purpose of this book is to address the second reason. I will inform you about a myriad number of ingredients that wind up in fast food, many of which are difficult to pronounce, disguised by acronyms, generically listed, or require a chemistry degree to understand their purpose.

Just to whet your appetite, here is a shortlist of mysterious chemicals that show up in fast food:

- **Azodicarbonamide**
- **DATEM**
- **Dimethylpolysiloxane**
- **Flavors**
- **Polyglycerol Esters of Fatty Esters**
- **Sodium Acid Pyrophosphate**
- **Sodium Stearoyl Lactylate**
- **TBHQ**

Here's the layout of the book:

- An introduction to the author and how he got interested in the ingredients found in junk foods, fast foods, and ultra-processed foods in general.

- A deep dive into selected menu items from McDonald's, Pizza Hut, and Taco Bell.

- Exploring research studies that investigate the connection between the consumption of ultra-processed foods and chronic diseases.

- Reviews of front-of-package food scoring systems from around the world to assist consumers in making healthy choices.

- Final summary of the book. Read this section if you want to skip all the details.

Let's get started!

Additional Information:

Fast Company, *The Forgotten History of the World's First Restaurant*

https://www.fastcompany.com/90669668/the-forgotten-history-of-the-worlds-first-restaurant

Contents

PREFACE... IV

PART I 1

INTRODUCTION – FROM ADDICT TO FANATIC.............................1

    CHAPTER 1.. 2

    THE FORMATIVE YEARS 2

    CHAPTER 2.. 4

    BEGINNING TO LOOK AT FOODS MORE CLOSELY 4

    CHAPTER 3.. 11

    MY REAL EDUCATION BEGINS 11

    CHAPTER 4.. 15

    INVESTIGATING ULTRA-PROCESSED FOODS 15

    CHAPTER 5.. 17

    FOCUSING ON FAST FOODS 17

    CHAPTER 6.. 19

    WHY READ THIS BOOK?... 19

        Your Body, Your Health *19*

        Fast Food Keeps Growing *21*

    CHAPTER 7.. 25

    WHAT'S IN THIS BOOK? 25

    CHAPTER 8.. 27

    WHAT'S NOT IN THIS BOOK? 27

    CHAPTER 9.. 31

    THE NITTY GRITTY... 31

        Selection of Fast-Food Restaurants to Survey *31*

        How To Use This Guide? *32*

        What Are Processed Foods? *39*

        More About the Processed Food Index (PFI).................... *42*

        Evaluating Fast Food from McDonald's, Pizza Hut, and Taco Bell ... *43*

        Some Takes on Highly Processed Foods *46*

        Examining Ingredients in Specific Menu Items *50*

    CHAPTER 10... 53

    KEY TAKEAWAYS IN PART I 53

PART II 55

EXAMINING FAST FOODS 55

    CHAPTER 11... 56

    INGREDIENTS IN SELECTED MCDONALD'S MENU ITEMS....... 56

        McDonald's Southwest Buttermilk Crispy Chicken Salad (500 Cal) .. *57*

McDonald's Quarter Pounder® with Cheese Bacon (620 Calories)....63

McDonald's Big Breakfast® with Hotcakes (1340 Calories)...............72

McDonald's McCafé® Mocha Frappé, Medium (510 Calories) 80

McD's McCafe´® Mango Pineapple Smoothie (250 Cal, Medium).....87

What About a Whole McDonald's Meal? 91

CHAPTER 12 94

INGREDIENTS IN SELECTED PIZZA HUT MENU ITEMS...........................94

Pizza Hut's Crispy Chicken Caesar Salad (830 Calories) 95

Pizza Hut's Tuscani® Creamy Chicken Alfredo Pasta (990 Calories) 107

Pizza Hut's Super Supreme Personal Pan Pizza® Slice (210 Cal) 113

P. H.'s Meat Lover's® Original Stuffed Crust® Slice (430 Calories) ...120

Pizza Hut's The Great Beyond Original Pan Pizza® Slice (390 Cal)...129

Pizza Hut's Cinnabon® Stick (80 Calories)......................................135

What About a Whole Pizza Hut Meal? 141

CHAPTER 13 146

INGREDIENTS IN SELECTED TACO BELL MENU ITEMS 146

Taco Bell's Power Menu Bowl - Steak (480 Calories) 147

Taco Bell's Quesarito - Steak (630 Calories)...................................155

Taco Bell's Grande Nachos - Steak (1080 Calories)..........................161

Taco Bell's Wild Strawberry Freeze™ (190 Calories for 20oz) 168

What About a Whole Taco Bell Meal?...174

CHAPTER 14 178

KEY TAKEAWAYS IN PART II 178

PART III...**180**

HEALTH STUDIES & SCORING SYSTEMS **180**

CHAPTER 15 181

HEALTH STUDIES OF ULTRA-PROCESSED FOODS 181

CHAPTER 16 228

INTERNATIONAL SCORING SYSTEMS 228

CHAPTER 17 263

KEY TAKEAWAYS IN PART III 263

PART IV **267**

AND SO, IT ENDS **267**

CHAPTER 18: 268

FINAL THOUGHTS 268

APPENDIX A: **273**

CALCULATING THE PROCESSED FOOD INDEX (PFI).............................**273**

APPENDIX B:...**278**

TABLE OF MCDONALD'S FAST FOOD WITH KEY DATA 278

APPENDIX C: 284

TABLE OF PIZZA HUT'S FAST FOOD WITH KEY DATA............................284

APPENDIX D 295

TABLE OF TACO BELL'S FAST FOOD WITH KEY DATA 295

APPENDIX E...300

FOOD PROCESSING TERMS & DEFINITIONS 300

APPENDIX F...303

GLOSSARY OF FOOD INGREDIENTS & ADDITIVES 303

ABOUT THE AUTHOR...349

ACKNOWLEDGEMENTS 350

INDEX...352

PART I

INTRODUCTION – FROM ADDICT TO FANATIC

CHAPTER 1

THE FORMATIVE YEARS

"To wash down your chicken nuggets with virtually any soft drink in the supermarket is to have some corn with your corn. Since the 1980s virtually all the sodas and most of the fruit drinks sold in the supermarket have been sweetened with high-fructose corn syrup (HFCS) -- after water, corn sweetener is their principal ingredient."

--- Michael Pollan, The Omnivore's Dilemma: A Natural History of Four Meals

Hey foodeaters, I have a confession to make. Once upon a time I was a junk food and fast-food addict. Yes, as a child of the 1950s and 1960s, I got "weaned" on convenience foods. Sunday evenings in front of the TV with my family were ritualized with "freshly" baked TV dinners served on handy, foldable TV trays. No, we didn't eat like that every day. There were many nights with home-cooked meals, but, over time, the variety of frozen entrees appearing in grocery stores kept increasing and many of them wound up in our freezer. Besides the all-in-one dinners from Swanson and Banquet which included main course, side, and dessert, there were frozen fish sticks, pot pies, macaroni & cheese, and Sara Lee desserts. Of course, as a kid, I didn't question what I was eating. If it tasted good, I was fine with it. If it didn't taste good, and I was hungry enough, I still ate it. Ingredients? Who cared about those? Maybe some manufacturers listed them on the packages, but who looked? And, of course, Nutrition Facts labels were nonexistent at that time.

Fast food restaurants first appeared to my life in the late 1960s. I grew up in the Age of Fast Food, at a time when many of the now iconic restaurants were just coming on the scene: Kentucky Fried Chicken and Churches Chicken (1952), Burger King and Shakey's (1954), Pizza Hut (1958), Domino's (1961), Taco Bell (1962), and Wendy's (1969). I can still remember the first McDonald's restaurant that opened in the suburb where I lived outside Dayton, Ohio. The French-fry smells wafting around the building were intoxicating and enticing. This was a restaurant that a teenage budget could afford and the food was fast and delicious. It was a great place to take a date, not only because it was affordable, but also, we could stay there as long as we wanted. And then there was the drive-up … how convenient was that? Even today, I get the concept and understand why people continue to flock to fast-food restaurants. Just like frozen dinners, I never questioned what I was eating in those restaurants as regards the ingredients or nutritional quality of the food. My wallet and palate determined my food choices back then. After the McDonald's store opened, many other franchises appeared in and around my city. The choices became dizzying and intoxicating. Thus began a quarter-life addiction to fast food.

Note to Readers:

This book does not have a bibliography. Instead, references are included in text boxes within each chapter, so the reader can more easily locate the references while reading the book and, thereby, immediately open a hyperlink (if using an electronic version) to take them to the website of interest. Of course, hyperlinks are not eternally stable, so, although all the links worked at the time of the writing of the book, some hyperlinks may be dead by the time you read the book. Often changed links can be found by simply Googling the title of the reference.

CHAPTER 2

BEGINNING TO LOOK AT FOODS MORE CLOSELY

"Fast food is popular because it's convenient, it's cheap, and it tastes good. But the real cost of eating fast food never appears on the menu."

— Eric Schlosser, Fast Food Nation: Fast Food's Impact on Society

A diet of mainly over-processed and fast foods dominated my teenage years and well into my twenties. During that time, I was lucky not to suffer from obesity or diet-related illnesses, but I was certainly setting myself up for an unhealthy future. Regarding the relationship between personal health and diet, I got a big clue while in college in the late 1960's. I discovered that I had gastrointestinal issues with milk and milk-containing products like ice cream. As I later learned, the condition I had was lactose intolerance, but that condition was not commonly discussed back then. Consequently, I stopped drinking milk, but I wasn't about to stop consuming some of my favorite foods, like dairy-laden desserts. It took another 10 years of intestinal distress to figure out the extent of my worsening problem and that I should stay away from lactose-containing foods. During the 1970s, I received BS and MS degrees in chemistry and later got interested in teaching chemistry at a community college. In my first teaching job, I had the opportunity to introduce a new course called "Chemistry for the Consumer," a survey course for non-science majors that would help fulfill a student's science requirement. This was not a hard-core chemistry class, but a practical guide to understanding the workings of consumer products using basic chemical concepts. It was this textbook that opened my eyes and raised questions about the foods I was eating. There were chapters on food chemistry with explanations about the use of additives in food products. After that, I started to pay more attention to what I was eating, but, nevertheless, I was still a junk food junkie.

After 10 years of teaching low-level chemistry and math classes, I changed careers. In the 1990s, I was hired by a global company that manufactured industrial and food ingredients (Tate & Lyle), most of which were derived from corn and soybeans. I worked as an analytical chemist, where I qualitatively and quantitatively tested the company's numerous products, creating new methods of testing, and assisted research chemists in developing novel products and applications to garner new business from the company's many commercial customers. My real education in the food ingredient industry began at that time. Until then, I had no clue about the enormous efforts that went into creating new food ingredients. As I learned, those innovations involved multiple years of development and testing conducted by food technicians, application chemists, analytical chemists, and pilot plant technicians and engineers. To bring a new product to market costs millions of dollars, whether the product was successfully marketed or not. From my memory, there were more losers than winners. What really boggled my mind was learning about all the products and food ingredients that could be produced from commodity crops like soy and corn. For example, field corn could be broken down into oil, starch, fiber, and protein components (gluten products). The starch could be deconstructed to produce numerous products like corn syrups, dextrose powder, fructose powder, and maltodextrins or it could be chemically and/or physically changed to produce numerous modified starches. The corn syrups could be chemically manipulated to produce still more products, like the omnipresent junk food ingredient, high fructose corn syrup (HFCS). And those products were just the conventional ones. The real money-makers that the company sold were the value-added ones, in which entirely new food ingredients were invented that provided unique properties for their customer's formulations. This is an on-going process in food manufacturing companies, which, as we'll see, has led to a plethora of new food additives showing up on product labels. Why, you ask? Two reasons come to mind: (1) The base commodities, like corn and soybeans, are abundant and cheap in the USA, and (2) Converting those cheap foodstuffs into something more valuable is very profitable for a food ingredient company.

At this point in my personal story, I want to take a side trip to bring up some examples of industrial ingredients that entered the marketplace to fulfill consumer demands. Hopefully, by the end of this section, it will be clear why these examples are significant. If you're of a certain age, you might remember the federal government's fat-intake warnings of the mid-1970s followed by the fat hysteria of the 1990s. In 1976, rising obesity rates in the USA were on the radar, and Senator George McGovern of South Dakota (later a presidential candidate) held a hearing of the U.S. Senate Select Committee on Nutrition and Human Needs to raise concerns about the overconsumption of fat and associated health issues, such as heart disease. In the following year, a document called the Dietary Goals for the United States was released. This document was the forerunner of the first Dietary Guidelines for Americans issued by the United States Department of Agriculture (USDA) in 1980. In short, the Dietary Goals recommended a 20% increase in the consumption of complex carbohydrates (e.g., starches) and naturally occurring sugars (e.g., found in fruits), a 45% reduction in refined sugars, and a 40% reduction in fat consumption. By the 1990s, the concerns over growing obesity rates were headlining newspapers, magazines, and TV news shows. Food manufacturers, seeing a ripe opportunity to address the problem, started to research ways to reduce the fat content of their foods, particularly in snacks and baked goods. Since fats and oils provided some key organoleptic (sensory) properties in foods, the manufacturers could not, in most cases, just pull the fat out of their products and replace it with something else, like carbohydrates.

However, there was one notable and delicious exception: Nabisco's Snackwell cookie cakes which were introduced in 1992 … these cookies were food engineering marvels, while also scoring with a great name! The chocolate Devils Food cookie, coming in at 50 calories per cookie, was simply to die for. Nobody, really nobody, missed the fat that was removed. Although a bit pricey, they were very popular and often sold out, but there was an annoying downside. People, including myself, tended to eat larger helpings of those cookies and, consequently, got bigger doses of calories. If you look at the ingredients, they were loaded with extra carbs, starting with the #1 ingredient, sugar, followed by high fructose corn syrup, corn syrup, and cornstarch. These cookies were touted as healthy, fat-free alternatives, but Nabisco didn't follow the recommendation of the Dietary Goals by also reducing refined sugars, rather, the unhealthy sugars were increased. The marketing was brilliant, consumers thought they were getting away with calorie theft, and the cookies garnered big profits for Nabisco.

Other food companies, like the ones making chips, crackers, baked goods, etc. got on the band wagon but they followed a different path. Instead, they looked for fat substitutes: new food ingredients that would mimic the properties of fat but eliminate or greatly reduce the fat load. So, a host of fat substitutes were devised, tested, and incorporated into conventional snack foods to see if they worked. The Food & Drug Administration (FDA) approved some of them including Simplesse, Leanesse, Salatrim, Enova Oil, Novagel, various gums, and Olestra. The latter ingredient has an interesting history. In 1968, two food scientists working for Procter

& Gamble (P&G), accidently discovered a chemical derivative of sucrose (table sugar) called a sucrose polyester that could mimic the properties of natural fat, but was itself noncaloric (zero calories). As a brand name it was called Olean (another smart moniker!). This synthetic oil could even be used for frying! Here was the answer for reducing fat in deep fried snacks, like potato chips. When somebody ate a food with Olestra in it, the fatty component would neither get absorbed or digested in the gut. Sounds like a miracle ingredient, right? It was touted as the single most important development in the history of the food industry, and was predicted to make Proctor & Gamble a tidy profit. However, getting approval from the Food & Drug Administration (FDA) was an uphill battle due to several health challenges particularly around digestive issues. But it finally got certified in January 1996 for use in savory snacks. However, the FDA did require a health notice on any food packages that contained Olestra which read: "Olestra may cause abdominal cramping and loose stools. Olestra inhibits the absorption of some vitamins and other nutrients. Vitamins A, D, E, and K have been added." Not the kind of warning a food manufacturer likes to see on their products. Eventually in 1998, Frito-Lay released a fat-free product called Wow chips, and at the same time, Proctor & Gamble started selling fat-free Pringles in a limited number of states. Surprisingly, the new product launches turned out to be a disaster for Frito-Lay. Consumers started reporting adverse health effects including severe diarrhea, fecal incontinence, and abdominal cramps. Sales plummeted. Proctor & Gamble fought like crazy, spending millions of dollars, to prove that Olestra was safe, but too many complaints had flooded into the federal government. Interestingly, the FDA never pulled its certification of Olestra, but the once-promising ingredient died a slow death from all the public outcries against it. Today, you will only find P&G selling reduced-fat Pringles made using conventional cooking oils and definitely without Olestra in them.

The second example comes from the company I worked for. As stated earlier, food companies were quick to jump on the band wagon of fat substitutes. With the new trend in the 1990s of removing or lowering the fat content of manufactured foods, there was money to be made. Since corn starch was cheap and readily available, a major research effort was conducted in the early 1990s to modify corn starch to produce a fat-like analog that could be used in dairy products, baked goods, fillings, and sauces. After several years of research, a prototype was devised that met the qualifications. The product was tested for efficacy by incorporating it in conventional dairy products, shared with potential customers, and scaled up for production. A portion of an existing plant was re-engineered and re-tooled to produce the new ingredient, a dry powder, on an industrial scale. Unfortunately, since the starch-based fat replacer had to be specially processed at the customer's manufacturing facility, using additional equipment to make a creamy additive, it was not a big seller. Despite further research to come up with an instant version, the new product never really caught on with food producers. After years of intensive research, labor, and millions of dollars, its production was eventually scrubbed.

So, what do we make of all this info? There are over 9000 certified food additives in the USA. The list is always growing. New ones get approved every year. Most of the new additives are synthetic, never having existed before. Why do food ingredient companies invest vast amounts of time and money to create new additives? First, if they can take a cheap ingredient and convert it into a functional ingredient that satisfies a demand in the food industry, the companies can make beaucoup bucks. Second, if a new diet craze comes along, food companies will devise new products to meet the demands of that diet. Think Atkins, paleo, keto, plant-based, and many others. As an example, just Google medium-chain triglycerides (MCTs), which never used to appear on food labels before the keto era. Now it shows up in many keto products. Third, the USA is a major influencer in the world. If a new, synthetic food ingredient becomes a hit here, it has a high probability of getting exported to many other countries, further improving the bottom line. As you'll see, the primary purpose of this book is to reveal the multitude of food additives that wind up in fast foods, and I think it's critically important to have some awareness as to why these additives are pervasive in our food supply and why their use continually expands.

Further Readings:
Dietary Guidelines for Americans, *History of the Dietary Guidelines*

https://tinyurl.com/2a3m8rh3

Center for Science in the Public Interest, *CSPI Warns Consumers About Frito-Lay 'Light' Chips with Olestra*

https://cspinet.org/new/200410251.html

United Kingdom, Ingredients Guide: *What Is in Our Food?*

http://www.food-info.net/uk/glossary/a.htm

Encyclopedia of Junk Food and Fast Food by Andrew F. Smith

https://tinyurl.com/4ts3mh7f

CHAPTER 3

MY REAL EDUCATION BEGINS

"The way we eat has changed more in the last 50 years than in the previous 10,000."

— Robert Kenner, filmmaker,
from the 2008 movie "Food, Inc."

Back to my personal story. By the early 1980s, after figuring out that I had the condition called lactose intolerance, I started to read food labels more carefully to avoid exposure to milk ingredients. Fortunately, for me, that was not a problem, because in 1966 the US government passed the Fair Packaging and Labeling Act, which required all consumer food products involved in interstate commerce to follow labeling guidelines. A subsection of the law required manufacturers to list ingredients on the label according to prominence, i.e., weight or volume. So, if you look at the ingredients panel for Dannon's creamy strawberry yogurt, here is what you'll see:

Ingredients: *Cultured grade A non-fat milk, cane sugar, water, corn starch, strawberry puree, contains less than 1% of agar agar, natural flavors, fruit and vegetable juice (for color), carob bean gum, lemon juice concentrate, vitamin D3.*

The first ingredient, non-fat milk, is present in the largest amount. The last ingredient, vitamin D3, is present in the smallest amount. The amounts of ingredients between the milk and the strawberry puree decrease accordingly. The manufacturers are not required to tell consumers the actual amounts of the ingredients, since that would give away a competitive edge, but sometimes hints are revealed. Notice that after "strawberry puree," the words "less than 1%" show up. We can interpret that to mean that the previous 5 ingredients add together to more than 93% of the product, but that the last 7 ingredients are each present at less than 1% by weight. (e.g., agar agar is less than 1% by weight, flavors are less than 1%, etc.) Note that for the last 7 ingredients, the descending order rule does not apply.

Assisted by the Labeling Act, I could determine for myself which food products contained milk and milk products, then choose to avoid them. Of course, that plan only worked if I had sufficient self-control when it came to foods I used to crave, like ice creams, shakes, and malts. That's another story!

Sadly, I wasn't paying much attention, other than curiosity, to all the other ingredients in processed foods back in the 1980s. And I don't remember other people caring about them either. However, with the advent of major public health concerns over chronic lifestyle diseases like obesity, diabetes, and cardiovascular disease, the federal government got interested in the nutritive quality of the foods that Americans were eating. In 1990, the Nutrition Labeling and Education Act was passed which required that all packaged foods carry nutrition labeling particularly around health claims. This law was enhanced in 1994 with the creation of the Nutrition Facts panel, which we all know and love today. Additionally, food labels were required to list the most important nutrients in a standard, easy-to-read format. Now, consumers could see how many calories they were consuming per serving, how many servings there were, how much fat, protein, and carbohydrates were in a serving, plus several other key nutrients like minerals, vitamins, and fiber. In one fell swoop, Americans had a whole bunch of new information on food labels about the nutritional quality of the food they were eating.

In the 1990s, I paid closer attention to the Nutrition Facts panel then I did to the ingredients listing, because I was more concerned about calories, salt, saturated fat, and other things that the medical experts were suggesting we should be tracking. At the same time, considering the work I was doing as an analytical chemist, I became more and more interested in the ingredients list, particularly why they were in the food. In parallel, as I made my way through my 40s, I linked my interest in nutrition with the food ingredients I was encountering in my workplace and wondered whether they might be adversely affecting my health.

So, for 20+ years I had a close-up look at hundreds of new and old food ingredients. I was involved in identifying and quantifying those ingredients, and, at the same time, I was learning about their functions and purposes. I got a privileged look behind the processed food curtain. I had an opportunity to explore the fascinating but, somewhat scary, world of food science. It was a world where real food could be manipulated, rearranged, and deconstructed using physical and chemical processes that would render the original food unrecognizable. The food industry and its supporting players, the food ingredient companies, were creating synthetic foods that never existed before but which provided exemplary texture, taste, smell, eye appeal, and shelf stability ... they were miracles of food science but were they good for us?

Speaking of shelf stability, I recall a little side experiment that I did while working as an industrial chemist. Well into my thirties, on my list of junk food favorites was the classic sponge cake with cream filling called the Twinkie. I couldn't get enough of them until I read a book called **Twinkie, Deconstructed** by Steve Ettlinger. The book traces the origins of the 35+ ingredients in that snack. Much later, when YouTube came along, I remember watching several videos about the apparent indestructibility of a Twinkie. They were rumored to have an indefinite shelf life due to chemical preservatives. So, I decided to test that hypothesis by placing a

Twinkie in an enclosed chamber and starting the clock. At the first signs of decomposition or mold formation, I would end the experiment. Day after day went by without a change. The trial continued for several weeks, when suddenly it came to an end. One day, my division manager was strolling through my lab, stopped in front of the Twinkie experiment, and asked curiously why a Twinkie was sitting on the lab bench. I sheepishly confessed to the experiment in progress; he wasn't impressed and pointed out that a very slowly rotting Twinkie wasn't a good look for the lab when visitors came through. So, in the name of science, my experiment to investigate the longevity of a Twinkie was relatively short-lived.

Further Reading/Info:
Twinkie, Deconstructed by Steve Ettlinger

https://tinyurl.com/racdc43j

YouTube: *This Twinkie Has Lasted 43 Years*

https://tinyurl.com/3dknbzdz

CHAPTER 4

INVESTIGATING ULTRA-PROCESSED FOODS

"America's most dangerous export was, is and always will be our fast-food outlets."

— Anthony Bourdain, Chef

After 20+ years working for a food ingredients company, I retired early to pursue my own interests. I started thinking about all my experience in chemistry education and industrial chemistry, as well as knowledge of food additives, so why not share my knowledge of ultra-processed foods with the world? That was in 2012. At that time, the new Internet phenomenon, podcasting, had been around just a short time, and I thought how cool would it be if I could put together a home-based podcast to share knowledge of the processed food industry. After one feeble attempt that was quickly abandoned, I placed the idea on hold for 4 years. Then, in 2016, I dusted off the idea, and using mainly YouTube videos and blogging sites, figured how to start a podcast. The podcast "Food Labels Revealed" was born later that year. Finally, I had a platform to inform the public about important issues in the commercial food industry. As far as I could tell, it was the only podcast of its kind dedicated to shedding "some light onto the sometimes mysterious and inscrutable food ingredients label, which are found on all processed food containers. What are all those chemicals in our packaged foods? Why are they there? Can they be harmful to our health?" Every month I posted a show where I ranted and raved about the processed food industry. I examined numerous grocery store food items, and described the food ingredients and nutritional properties listed on the Nutrition Facts panels. Sometimes, I went into mind-numbing detail. From my perspective, it was time to give back to society and share what I knew about the processed food industry. Maybe sharing that knowledge would help some people to think more deeply about what they were putting in their bodies. I would try to share whatever expertise I had in that field with consumers, and try to explain facets of food chemistry, food additives and associated nutritional facts in as simple and clear a manner as possible. The podcast developed a small, dedicated but slowly growing following over time. Eventually, I realized that there was just a select sub-group of people who were interested in knowing what was in ultra-processed foods. I got a number of very positive and appreciative messages from those folks. However, I learned that most people, even my friends and relatives, weren't interested in my message. As far as food was concerned, they liked what they liked. They didn't want to hear bad news about food ingredients. The old expression, "ignorance is bliss," became obvious. As I learned over time, most people would rather hear good news about their bad habits ... not the opposite. I certainly understand that attitude. We live in a world where we are constantly bombarded with bad news. Despite the lackluster reception to the Food Labels Revealed podcast, I still thought it was essential to deliver the message.

Additional Information:
Hosting website for the Food Labels Revealed podcast

http://foodlabelsrevealed.podbean.com/?source=pb

CHAPTER 5

FOCUSING ON FAST FOODS

"In twentieth-century Old Earth, a fast-food chain took dead cow meat, fried it in grease, added carcinogens, wrapped it in petroleum-based foam, and sold nine hundred billion units. Human beings. Go figure."

— Dan Simmons, author, "Hyperion"

After examining and reporting on various commercial foods found in grocery stores, in 2018, I turned my attention to fast foods. Were the same food additives found in sauces, dressings, cereals, kid's lunches, frozen foods, breads, quick-fix meals, etc. also found in fast foods? Of course, the problem with that question is the fact that you don't get a Nutrition Facts panel with your hamburger or fried chicken takeout, so how would you know what ingredients were used? And, I couldn't just walk into a fast-food restaurant and demand a list of ingredients. Fortunately, the FDA in 2018, as a result of alarming reports about obesity rates in the USA and associated increases in Type 2 diabetes, required that fast food restaurants start posting nutritional data about their products. This information is now available in all chain stores, but it's also readily found at company websites. So, I began looking for the information I needed online. Lucky for me, there were a handful of companies that also provided full or partial ingredient lists for their menus, similar to the information found on commercial food labels. This data was voluntarily provided at the company's website which was not mandated by any federal regulatory agency. Any consumer with an Internet connection can access this information, although, in some cases it may seem hidden in plain sight. With the necessary information at hand, it was easy to get answers to my questions about additives used in fast food: (1) Is there really a difference between fast foods and the types of highly processed foods found in the typical grocery store? (2) Is fast food intrinsically unhealthy? and (3) How similar are the ingredients in fast food meals across different restaurants? After an evaluation of a half dozen or so fast-food restaurants, it became evident to me that those restaurants were using the same additives in their menu items as the major processed food manufacturers used in their grocery store products. I recorded 6 podcast episodes on fast foods.

That knowledge spurred the writing of this book, which, as far as I know, is the first of its kind. I decided to provide a detailed accounting of the ingredients used in 3 major, international fast-food restaurants. There are no secrets to reveal. The data was publicly posted at the websites of the restaurants during the writing of the book. The main purpose is to compile the information in one place, explain and comment on the purposes behind the ingredients based on my chemistry background and other sources, to attempt to rate the degree of processing of fast foods, and to discuss recent scientific studies revealing health problems associated with the consumption of fast foods.

Additional Reading:
Hosting website for the Food Labels Revealed podcast
http://foodlabelsrevealed.podbean.com/?source=pb
FLR 027: Burger Breakdown – McD's vs. BK
https://tinyurl.com/h9e93j9x

CHAPTER 6

WHY READ THIS BOOK?

"If we can have a fast-food restaurant on almost every corner, then we can certainly have a garden."

--- Bill de Blasio, former mayor of NYC

Your Body, Your Health

Here's a simple answer to the title question. If you eat fast food from the thousands of restaurants strewn across the United States, you should read this book. If you have even a passive concern about what you put into your body when you eat fast food, you should read this book. If you are a fast-food eater and suffer from chronic disorders like diabetes, obesity, high blood pressure, kidney disease, cardiovascular diseases, intestinal problems, and cancers, you should read this book. I cannot claim with 100% proof that all your health problems are strictly due to eating fast food or highly processed food in general, but there is mounting epidemiolocal evidence, that, diets rich in highly processed foods, contribute to the decline in health over the course of a lifetime. I'll showcase some notable research studies in Part III. Of course, we can't blame the fast-food industry for all the health problems of the American population because health is a complicated matter involving many factors including heredity, amount and frequency of exercise, mental health, poverty, availability of healthy food, etc., etc. But there are certainly strong correlations between fast food consumption and negative health outcomes. In recent years, the governments of other nations (sadly, not so much in the USA), are actively seeking to minimize the impact of highly processed foods on their populations, particularly in children, who, for the most part, don't have the education, independence, or self-control to understand or withstand the tremendous marketing influences of fast food and junk food manufacturers. Plus, people in European countries have access to what's called an E-number index which identifies food additives by a specific number and consumers can easily locate information about those additives using a smart phone or other computer device. More about that later.

Additional Information:
Complete List of ENumbers

https://tinyurl.com/2ar3c5fy

Non-Food in Your Food

Food companies have available to them over 9,000 additives that the federal government, under the auspices of the Food & Drug Administration (FDA), certifies for use in commercial products (See reference below, "Substances Added to Food"). The list grows every year as new additives are researched and submitted to the FDA for approval. It would be fantastic if the FDA, like the European Union, provided a single website listing all additives, descriptions, and health information, but if such a reference exists, I'm not aware of it.

In almost all cases, new additives fall under the classification GRAS, which stands for Generally Recognized as Safe. Although new food additives are required to be safe for consumption, the federal government rarely performs investigative studies to determine their level of safety. That job is left to the manufacturers or their designees (contractual labs). A food manufacturer submits their petition, including test data, to the FDA and after a review by

FDA scientists, receives a notice of approval, limited use, or a rejection. The vast majority of petitions are approved. The only time the FDA gets involved with a safety investigation is when a health problem arises after the fact, i.e., following the introduction into the food supply. There are many examples of after-the-fact problems (See reference below, "Safety Rankings for Food Additives"). I'll name just a few here: (1) Partially hydrogenated fats and oils, artificial substances banned in 2015, showed evidence of a link to heart disease due to the presence of trans fats in them (they were in the food supply for decades in margarines and spreads); (2) Violet 1, an artificial coloring agent used to stamp USDA-approved beef, was banned in 1973 as a suspected carcinogen; (3) Ethyl Acrylate, an artificial fruity flavoring, was de-GRASed in 2018 as a potential cancer-causing agent; and (4) cyclamate, an artificial sweetener, was stripped of its GRAS status in 1969 as it amplifies the effects of cancer-causing substances. So, should we blindly trust that all the current FDA-approved additives in our food will be safe indefinitely? That's a rhetorical question, but I would answer "No."

Further Readings:
Food & Drug Administration: *Substances Added to Food*

https://tinyurl.com/2uedc4zv

Center for Science in the Public Interest: *Chemical Cuisine Ratings*

https://tinyurl.com/3x3y4c3n

Fast Food Keeps Growing

If you live in a city of decent size, you will likely be within a few minutes driving distance of dozens of fast-food restaurants. That's no accident. Fast food purveyors are well aware of the trends in fast food consumption in the America, and they spread their tendrils far and wide. The trend has seen a steep rise since the 1950s. Just think for a moment of the fast-food restaurants that exist today that weren't around in your childhood. And that trend continues. New restaurant chains are popping up all the time. Here is a shortlist of newbees that have recently emerged on the fast-food stage: Mici (Italian), Pokéworks, Teriyaki Madness, and Maui Tacos. They may be coming to your neighborhood soon! My point is that fast food has been with us a long time, it is here to stay, and the types of fast-food cuisines will continue to expand, particularly in the ethnic sector.

Let's look at some stats provided by the federal government, namely the Centers for Disease Control (CDC). This data comes from their branch, the National Center for Health Statistics and covers the years 2013 to 2016.

- 37% of adults (~85 million) consumed fast food on a given day.

- As expected, younger adults (20 to 39 years) consumed more fast food at 45%.

- The segment of the population consuming the most fast-food was non-Hispanic blacks.

- The percentage went up as the income level went up indicating that poverty is not the cause of fast-food consumption.

- About 11.3% of total daily calories came from fast-food.

The following information was provided by the United States Department of Agriculture (USDA), Economic Research Service (ERS):

- In the years 2009 to 2014, the number of fast-food restaurants grew by 9%. In 2014, the total number was 228,677.

- The states that saw the largest growth were Connecticut, North Dakota, New York, and Texas. In those states, the restaurants grew at a faster pace than the populations.

- Overall, number of restaurants per person has increased.

- The fastest growing county in the USA was Los Angeles, which increased by 10% (680 new restaurants).

The Washington Post makes some interesting points about the American diet in a June 2014 article.

- Americans on average now eat nearly 2,600 calories a day, almost 500 more than they did forty years ago. [Note: Federal guidelines recommend 2,000 calories for the average American.]

- Over 92% of the uptick in per capita caloric intake since 1970 is attributable to oils, fats, and grains. [Estimated at 47% of calories consumed].

- The two food groups that Americans are eating more and more of, added fats/oils, and flour/cereal products, are the ones commonly found in most processed and fast foods.

- Since the 1970s, fast food has become a significant part of the American diet. Between 1977 and 1978, fast food accounted for just over 3% of calories in the American diet, but, between 2005 and 2008, that amount leaped to over 13%.

- If you are an American over 20, you are 3 times more likely to be obese than people 30 years ago.

The conclusion? A significant amount of the food that Americans consume is coming from fast-food restaurants, and we know, from numerous studies, that fast food is associated with nutritional deficiencies, weight gain, risk of obesity, and chronic lifestyle diseases. BUT, on top of those issues is the fact that consumers don't know what they're eating. Yes, they know the menu items and their basic components, like bacon, lettuce, tomato, and mayonnaise in a BLT., and, if they pay attention, they know the nutritional values of the food, but they don't know what's in the food: the **actual** ingredients. That's the purpose of this book.

Further Readings:
Centers for Disease Control: National Center for Health Statistics

"Fast Food Consumption Among Adults"

https://tinyurl.com/3da3d62r

United States Department of Agriculture: Economic Research Service

"Higher Incomes and Greater Time Constraints Leads to Purchasing More Convenience Foods"

https://tinyurl.com/26vlhgml

Washington Post, *"How the American Diet Has Failed"* by Roberto Ferdman

https://tinyurl.com/bw3yww39

CHAPTER 7

WHAT'S IN THIS BOOK?

"I went into the library and read about fast food and became amazed by all the stuff I didn't know. I learned that there is a whole world behind the counter that, it seemed to me, has been deliberately hidden from the public."

— Eric Schlosser, author

I have selected a shortlist of three iconic, fast-food restaurants and collated the ingredients found in their menu items. For each restaurant, a document has been created providing (1) an ingredient list for every menu item for which that information is available, (2) a list of how many ingredients fall within each processing category from unprocessed to extremely processed, and (3) a calculated value for the overall degree of processing for the food item. The massive amount of data collected for this book is available at several Google hosting sites mentioned later. In subsequent sections of the book, I address a number of menu examples from the three restaurants.

Since very few readers are likely to be chemists or food scientists, Appendix E provides a list of food processing terms and definitions which will help explain the various functions of food additives.

Appendix F provides a glossary of food ingredients and additives. This is an important reference section. All of the significantly processed ingredients found in the menu items are listed alphabetically for easy lookup, plus informative descriptions are provided. This glossary can serve as a guide to consumers who are interested in identifying industrial foods and additives routinely used in fast foods. Although I have only surveyed 3 fast-food restaurants out of literally hundreds of possibilities, my experience has shown that the types of ingredients and additives used by these restaurants are common across the entire fast-food industry. However, I don't claim that this glossary is comprehensive and complete, but it probably captures the majority of significantly processed ingredients and industrialized chemicals used by fast-food restaurants.

A frequent criticism about comparing natural foods to industrial foods involves composition. Some critics say that, e.g., coffee, has hundreds of individual constituents in it. Actually, there are over 1000 chemical compounds in coffee. Isn't that true of fast foods? Are they not just simply combinations of large numbers of compounds? Yes, that's true, but a key fundamental concept is missing from this comparison. There are two considerations. First, natural foods have developed (evolved) over tens of thousands of years if not longer. Humans have learned, through trial and error, which foods are edible and healthy and which ones are inedible and cause sickness or death. The synthetic food components invented by humans have only been around for a few hundred years. Second, food additives are prepared and tested in isolation, not combined together as in natural foods over the course of human evolution. No one really knows the consequences of the synergistic effects of foods loaded with synthetic additives. For example, if four additives with questionable health effects are combined, is the sum effect worse than the individual components by themselves? Rarely do scientists study combinations of synthetic additives on human health. These reasons point out the fallacy of comparing the compositions of natural foods vs. industrial foods.

CHAPTER 8

WHAT'S NOT IN THIS BOOK?

"We do not subsidize organic food. We subsidize these four crops - five altogether, but one is cotton - and these are the building blocks of fast food. One of the ways you democratize healthy food is you support healthy food."

— Michael Pollan, author

To clarify the purpose of this book, I want to establish from the start what are <u>not</u> my intended goals. The main objective is very specific as regards the reporting of ingredients and additives used in fast foods. However, this book is:

Not an Historical Account

I will not be discussing the development of fast foods, the history of fast-food restaurants, the economic impact of drive-through eating establishments in the United States, and the historical associations between public health and fast-food consumption. There are plenty of books that do a wonderful job addressing these subjects. I'll refer the reader to a couple of good ones.

Encyclopedia of Junk Food and Fast Food (2006) by Andrew F. Smith

Like the classic Encyclopedia Britannica, this book literally lists names and terms from A to Z. Pretty much anything you can imagine regarding junk food and fast food is reverenced in this book. It has many historical tidbits about the origins of restaurants and specific foods.

Fast Food and Junk Food: An Encyclopedia of What We Love to Eat (2011) by Andrew F. Smith

This book is a larger and updated version of the above. There are over 800 pages, so this book is broad in its scope.

Fast Food Nation: The Dark Side of the All-American Meal (2012) by Eric Schlosser

This book provides a history of the rise of fast food in American society and explores every aspect of the industry including cultural, technical, economic, and political aspects.

Further Readings:
Encyclopedia of Junk Food and Fast Food by Andrew F. Smith

https://tinyurl.com/4ts3mh7f

Fast Food and Junk Food: An Encyclopedia of What We Love to Eat by Andrew F. Smith

https://tinyurl.com/npv97kkm

Fast Food Nation: The Dark Side of the All-American Meal by Eric Schlosser

https://tinyurl.com/tzejr4d4

Not a Nutrition Guide

Although nutritional information for fast food restaurants is low hanging fruit (every restaurant is required to provide it) and certainly deserves attention, other than providing calorie counts for menu items, I won't be presenting data or commentary on that subject. That would be a major undertaking and definitely worth doing, so I hope that some motivated food scientist, dietician or nutritionist tackles that job. I have not come across any books strictly pertaining to that topic. The examination of nutritional data for fast foods is the key to linking the consumption of fast foods to poor health outcomes for Americans. I will touch upon that subject when I discuss recent research studies in Part III of the book.

Not a Food Review

For this book, I have gathered data on hundreds of fast-food menu items, but I won't be saying a word about food quality, appeal, prices, service, or popularity. Revealing and discussing ingredients is the centerpiece of this book.

Not a Commentary on Fast Food Cravings

It's not only the convenience, cheapness, ready availability, and taste that attract people to fast food, but many psychologists have investigated the brain chemistry responsible for food cravings and even food addictions. However, that subject will not be addressed in this book. If you want to look into the subject of fast-food cravings, here are some recommendations.

Further Readings:
The End of Overeating: Taking Control of the Insatiable American Appetite by David Kessler

https://tinyurl.com/sasn2z92

The Pleasure Trap: Mastering the Hidden Force that Undermines Health & Happiness by Douglas Lisle and Alan Goldhamer

https://tinyurl.com/488kech7

Hooked: Food, Free Will, and How the Food Giants Exploit Our Addictions by Michael Moss

https://tinyurl.com/kvenrfe6

Processed Food Addiction website (Dr. Joan Ifland)

https://pfa.mykajabi.com/

Not an Exposé on Hidden or Incidental Food Ingredients

The ingredients listed on a food label are not the only substances found in foods. Processed foods can be contaminated by residues from processing machinery, cleaning products, and chemicals used in plant and animal product production, e.g., pesticides, herbicides, antibiotics, veterinary medicines, etc. Also, insect and animal residues may be present. Yes, reader, there are actual legal limits to zoological contamination. For a quarter cup of cornmeal used to make tortillas, the federal Food & Drug Administration (FDA) allows one or more whole insects and 50 or more insect fragments. Also, the FDA allows two or more rodent hairs or one or more poop fragments. The FDA calls these unintended ingredients "defects." This subject is certainly worthy of investigation, but it will not be covered in this book.

CHAPTER 9

THE NITTY GRITTY

"Fast food may appear to be cheap food and, in the literal sense it often is, but that is because huge social and environmental costs are being excluded from the calculations."

— King Charles III, English monarch

Selection of Fast-Food Restaurants to Survey

Given that there are hundreds, if not thousands, of fast-food companies, it would be a monumental task to investigate all of them, particularly since the majority of them don't publicize ingredients used in their foods. Plus, given my experience in evaluating ultraprocessed and junk foods, I strongly suspected that the additives used in fast foods would be common throughout the industry. So, I decided to limit my attention to multinational, big-name restaurants that represented a good cross section of American fast food, and that also provided readily available online listings of menu ingredients. I chose three categories of fast food that most Americans could relate to, namely hamburgers, pizza, and Mexican food. The selected restaurants are McDonald's, Pizza Hut, and Taco Bell respectively. In the latter half of 2020 and the first half of 2021, I collated ingredient data for these restaurants. Note that the data collected was a snapshot in time, so the menu items listed in this book may have changed, been dropped, or new ones may have been added by the time you read this book. The following table lists the number of menu items that I evaluated for each restaurant.

Fast Food Restaurant	Number of Menu Items Examined
McDonald's	91
Pizza Hut	155
Taco Bell	74

Further Readings:

McDonald's

https://tinyurl.com/3xf8jyda

Pizza Hut

https://tinyurl.com/bdcmjkd3

Taco Bell

https://tinyurl.com/4fhefvxs

How To Use This Guide?

There are hundreds of food and beverage items offered by McDonald's, Pizza Hut, and Taco Bell. However, to list every ingredient for each restaurant and evaluate their degree of processing would be a huge task. So, I have chosen to include a select group of food items from each restaurant that are both popular and represent a broad range of additives. For those readers who want to see the full ingredient lists for each restaurant, the following Google links will take you to documents containing the complete data. There is a boatload of data provided for each restaurant. Additionally, at the end of this book, there is a Glossary (Appendix F), which lists the names and descriptions of all the ingredients that I have determined are moderately to extremely processed (see the section below for more details about processed food descriptors).

McDonald's: https://tinyurl.com/5acnks3r
Pizza Hut: https://tinyurl.com/z72em99m
Taco Bell: https://tinyurl.com/erdrpxba

For every food item evaluated, here is what you'll see:

- A text box showing the name of the food item and the calorie count if known.

- An ingredient list that conforms to the FDA regulation to begin with the ingredient that's present in the largest amount by weight followed by ingredients in descending order of amount. The ingredients are sequenced numerically for easy reference. All data originates from the restaurant's website.

- A Degree of Processing (DP) table that alphabetically lists significantly processed ingredients. Unfortunately, there is no definitive handbook or guide for determining the degree of processing of a food ingredient. The descriptors used in the DP table are assigned by the author based on personal knowledge and reasonable criteria. Only those ingredients that are designated as moderately to extremely processed will appear in the DP table. In general, the more an ingredient deviates from a natural food source, the more processed it is. Such industrial processes as separating parts of food, breaking down food structure, chemically or physically modifying the food, preferentially extracting a food component, chemical synthesis, etc. all contribute to classifying ingredients as moderately to extremely processed. I know there will be some people who will disagree with my classifications, but, until there is a standard reference, the classifications will remain a matter of informed opinion.

- For easy reference, each ingredient in the table has a number(s) by it indicating where it resides in the main ingredient list. If an ingredient is artificial (lab made), then it is designated by the word "synthetic."

- The next section provides a breakdown of the ingredients showing the number in each category of processing from "unprocessed" to "extremely processed." These terms will be discussed in the section entitled "What Are Processed Foods" below.

- The last section provides the total number of ingredients in the food item, the number of unique ingredients, and an author generated score called the Processed Food Index (PFI). If an ingredient is used more than once in a menu item, it is only counted once as a unique ingredient. The sum of unprocessed to ultraprocessed ingredients should equal the number of unique ingredients. The Processed Food Index is a numerical value ranging from 0 to 100 and represents the degree of industrialization. The higher the index value, the more industrialized is the food item.

Here is an example to illustrate the information that accompanies each food or beverage item in the book. Let's look at the basic, no-frills hamburger from McDonald's.

SECTION 1

ITEM NAME	CALORIE COUNT
Hamburger	250

BUN:

Flour (Wheat Flour(1), Malted Barley Flour(2), Niacin(3), Iron(4), Thiamine Mononitrate(5), Riboflavin(6), Folic Acid(7)), Water(8), Sugar(9), Yeast(10), Soybean Oil(11), Contains 2% or Less: Wheat Gluten(12), Potato Flour(13), May Contain One or More Dough Conditioners (DATEM(14), Ascorbic Acid(15), Mono and Diglycerides(16), Enzymes(17)), Vinegar(18).

BEEF:

100% Pure USDA Inspected Beef(19).

KETCHUP:

Tomato Concentrate from Red Ripe Tomatoes(20), Distilled Vinegar(21), High Fructose Corn Syrup(22), Corn Syrup(23), Water(24), Salt(25), Natural Flavors(26).

PICKLE SLICES:

Cucumbers(27), Water(28), Distilled Vinegar(29), Salt(30), Calcium Chloride(31), Alum(32), Potassium Sorbate (Preservative)(33), Natural Flavors(34), Polysorbate 80(35), Extractives of Turmeric (Color)(36).

ONIONS:

Chopped Onion(37).

MUSTARD:

Distilled Vinegar(38), Water(39), Mustard Seed(40), Salt(41), Turmeric(42), Paprika(43), Spice Extractive(44).

GRILL SEASONING:

Salt(45), Pepper(46).

Notice that each ingredient is given a number to get a total count. These numbers are also found in the "Degree of Processing Table" in Section 2. Note that the only nutritional data that is reported is the calorie content of the food or beverage.

SECTION 2

DEGREE OF PROCESSING (DP) TABLE

Ingredient ID (List Number)	How Processed?
Alum (synthetic) (32)	Extremely
Ascorbic Acid (synthetic) (15)	Extremely
Calcium Chloride (synthetic) (31)	Extremely
Corn Syrup (23)	Highly
DATEM (synthetic) (14)	Extremely
Enzymes (17)	Highly
Extractives of Turmeric (36)	Highly
Folic Acid (synthetic) (7)	Extremely
High Fructose Corn Syrup (synthetic) (22)	Highly
Malted Barley Flour (2)	Moderately
Mono and Diglycerides (synthetic) (16)	Extremely
Natural Flavors (26,34)	Moderately
Niacin (synthetic) (3)	Extremely
Polysorbate 80 (synthetic) (35)	Extremely
Potassium Sorbate (synthetic) (33)	Extremely
Potato Flour (13)	Moderately
Riboflavin (synthetic) (6)	Extremely
Soybean Oil (11)	Highly
Spice Extractives (44)	Highly
Sugar (9)	Moderately
Thiamine Mononitrate (synthetic) (5)	Extremely
Vinegar (18,21,29)	Moderately
Wheat Flour (1)	Moderately
Wheat Gluten (12)	Moderately

The ingredients in the table are listed alphabetically. Note that in Section 2 only the ingredients that are determined to be moderately to extremely processed are included in the table. Ingredients that are unprocessed or lightly processed are omitted since they are considered inconsequential contributors to the industrialized state of a food. Refer to the sub-section below entitled "What Are Processed Foods" to find more information about these designations. The ingredients that are listed as "synthetic" refer to artificial substances manufactured in chemical plants. In some cases, synthetic ingredients may have natural analogs, e.g., in the table above, ascorbic acid or vitamin C is found in many fruits and vegetables as a natural substance, but, for cost-saving reasons, food companies are likely to use the synthetic version.

***

SECTION 3

OF UNPROCESSED INGREDIENTS: 10
OF LIGHTLY PROCESSED INGREDIENTS: 2
OF MODERATELY PROCESSED INGREDIENTS: 7
OF HIGHLY PROCESSED INGREDIENTS: 6
OF EXTREMELY PROCESSED INGREDIENTS: 11

TOTAL # OF INGREDIENTS: 46
OF UNIQUE INGREDIENTS: 36 (78%)
PROCESSED FOOD INDEX: 43 (Highly Industrialized)

* See Appendix A to learn how the PFI was calculated.

***

The calculated Processed Food Index (PFI), shown above for this McDonald's menu item, determines how industrialized it is according to the table below.

Non-Industrialized:	0
Lightly Industrialized:	1 to 10
Moderately Industrialized:	11 to 40
Highly Industrialized:	41 to 70
Extremely Industrialized:	71 to 100

In Section 3, the ingredients are broken down into categories according to how processed they are. This hamburger has 10 unprocessed ingredients, 2 lightly processed, 7 moderately processed, 6 highly processed, and 11 extremely processed. Of the 46 total ingredients, 36 of them are unique and 10 are replicated. This means that if you were to make this hamburger in your kitchen from scratch (assuming you don't have pre-made pickles, mustard, and ketchup), you would need to gather up 36 ingredients. The "Processed Food Index" or PFI is a calculated number based on the number and type of processed ingredients. The higher the number of significantly processed ingredients the higher the PFI. I assign a descriptor for the food item based on this number according to the table shown above. In this case, the PFI score was 43, so the hamburger is described as "highly industrialized." See below for more details about the PFI scores.

What Are Processed Foods?

Let's be honest here. The word "processed" as applied to food can mean just about anything these days. The term has become generic for any food not found in its raw form. In recent times, the word has taken on a negative connotation as commercial food has become more and more refined, manipulated, altered, and chemicalized, such that it very faintly, if at all, resembles the natural ingredients from which it was derived.

Food processing has been with us, ever since humans realized that, in times of plenty, food needed to be preserved to protect against shortages, spoilage, and unpredictable disasters. Various ancient techniques were applied such as drying, smoking, fermenting, and pickling. Food processing, in and of itself, is often a help and not a hindrance to maintain food quality and availability. In modern times, industrialization introduced new techniques such as freezing, extracting, centrifuging, chemical preservation, irradiation, extrusion, and others. Also, foods increasingly became physically and chemically manipulated. Food additives changed color, altered taste, and performed other functions to totally change the look and appeal of food. These changes have led to reductions in food quality and potential threats to public health.

Unless you go out to your garden or a grocery produce sections, pick up a vegetable or fruit from a plant or tree, then chow down on it, pretty much any food that you buy in a store has been processed. What really makes the difference is the degree of processing. That degree of processing will determine the health value of the food. If you buy a head of lettuce, either whole or pre-packaged, it will, at least, have been washed, a very basic level of processing. One might even describe this food as unprocessed since its integrity was not changed. No one is likely to argue that this slight degree of processing reduces the quality of the lettuce. The nutrient value of the lettuce will remain intact.

Another example … corn. Let's say you're hankering for some corn-on-the-cob. You buy some fresh, sweet corn at a grocery store or farmer's market. First, you need to husk or shuck it to remove the leafy wrapping and silky tassel. Then, you'll probably want to wash it to remove dirt and insects. Since you're not likely to eat the corn raw, you steam, broil, or boil it. Maybe you'll coat the warm corn with some butter or margarine for flavoring. This food has definitely been processed, but its integrity and much of its nutritive value have been preserved. Now, let's take a look at ultra-processing. The FritoLay company makes a product called Cheetos® Crunchy Flamin' Hot® corn chips. The chief ingredient is processed corn. Here are the ingredients:

CHEETOS® Crunchy FLAMIN' HOT®

Enriched Corn Meal (Corn Meal, Ferrous Sulfate, Niacin, Thiamin Mononitrate, Riboflavin, Folic Acid), Vegetable Oil (Corn, Canola, and/or Sunflower Oil), Flamin' Hot Seasoning (Maltodextrin [made from corn], Salt, Sugar, Monosodium Glutamate, Yeast Extract, Citric Acid, Artificial Color [Red 40 Lake, Yellow 6 Lake, Yellow 6, Yellow 5], Sunflower Oil, Cheddar Cheese [Milk, Cheese Cultures, Salt, Enzymes], Onion Powder, Whey, Whey Protein Concentrate, Garlic Powder, Natural Flavor, Buttermilk, Sodium Diacetate, Disodium Inosinate, Disodium Guanylate), and Salt.

In this case, the corn is likely field corn (dent), not sweet corn. This is the most common form of corn grown in the United States, which blankets the Midwest and is the source of a multitude of industrial products, such as cornstarch, syrup, oil, animal feed, etc. It's processed in the following ways: (1) de-husked, (2) washed, (3) de-kerneled, (4) dried, and (5) ground into a powder to make corn meal. There are definitely more processing steps involved here compared to corn-on-the-cob, all of which involve industrial machinery. I would classify the corn meal as <u>moderately</u> processed. But to make Cheetos®, the corn meal gets fortified with vitamins and minerals, deep fried in oil, seasoned with herbs and chemicals, colored with artificial dyes, and coated with cheese powder, a mix of natural and synthetic substances. The final product doesn't resemble corn at all and contains a host of industrial additives, many of which the average consumer would not recognize, much less be able to pronounce. I categorize Cheetos® as a <u>highly</u> processed food, a food, unlike corn, that seemingly melts in your mouth, tastes nothing like corn-on-the-cob, and will last an incredibly long time in the original sealed bag. Note that Fritos® Original Corn Chips, a less industrialized product, available for around 90 years, has only three ingredients: corn, corn oil, and salt. I would label this simpler snack as <u>moderately</u> processed.

So, commercial foods are all processed in some way, but what distinguishes one from the other is the degree of processing, which can range from lightly processed to highly processed. Another term which has gotten quite popular in the media and in research articles is "ultra-processed." This is just another way to describe a highly processed food. However, as a descriptor, it lacks nuance. How do you go from lightly processed food to ultraprocessed food? Is it lightly to moderately to highly to ultraprocessed? These definitions are nebulous. That's why I prefer in this book to describe individual ingredients as unprocessed, lightly processed, moderately processed, highly processed, and extremely processed to capture the full range of possibilities. It's important to understand that there is no reference guide to assign each food ingredient a degree of processing. No such guide exists, so I have made these assignations based on my knowledge of food ingredients/additives. For sure, it's not a perfect system because it lacks rigor, but, at least, it provides a starting point to compare different types of commercial foods. After assigning descriptors to the ingredients/additives of a food product, then I calculate a Processed Food Index (PFI) based on the number of ingredients in each category. Finally, the PFI value is associated with a degree of industrialization. I use the word "industrialization" because to me it carries a clearer meaning than "processed." It stimulates a particular image in a consumer's mind. Foods that are industrialized are manufactured in factories and are not simply natural foods that are cleaned up or prepared for eating. This subject is explored further in the next section.

More About the Processed Food Index (PFI)

If you're curious about how the PFI is calculated, see Appendix A.

I prefer not to get bogged down with math in this section ... there's always the danger of losing too many readers!

The PFI provides a relative scale to compare commercial foods in terms of how industrialized they are. The PFI is not rigorously scientific and certainly not perfect, but it can be a useful tool for evaluating the "unnaturalness" of a processed food. By "unnaturalness" I mean how far the food product has been altered from its original state. The hypothesis is that the higher the PFI value, the lower the nutritional value, the unhealthier the food will be, and the less natural it is. This idea will be explored in more detail at the end of the book, where I discuss research studies that correlate the consumption of highly processed foods with negative health outcomes (See Part III).

However, the PFI does have a major downside. It doesn't work well with commercial foods that have only a handful of ingredients or less. The reason for this drawback is the lack of information about the amounts of ingredients listed on food labels or at fast-food restaurant websites. Recall that the ingredient list starts with the ingredient present in the largest amount and proceeds to list all other ingredients in decreasing order of amount. That's useful information, but it's very incomplete. Let's look at another example from the McDonald's menu:

Apple Slices

Ingredients: Apples(1), Calcium Absorbate(2).

There are only two ingredients in this food. The apples, of course, are the major ingredient in terms of amount and the calcium absorbate is the minor ingredient, but from the information provided, we have no idea how much of each ingredient is present in the food item. However, I do know that calcium absorbate, a relative of ascorbic acid (vitamin C), is used as a preservative. Generally speaking, preservatives are used in very low amounts, so the apples likely represent well over 95% of this food item. If I calculate the PFI, the result is 50, indicating that this food is highly industrialized. Based upon what I just stated, this conclusion is ridiculous, since the food item is mostly apples. If I knew the amount of each ingredient, the PFI result would have come out much lower with a classification of "lightly industrialized." But, unless you work for a food manufacturer, the data regarding ingredient amounts are not available. So, the lesson here is not to trust PFI values for food items that have only a small number of ingredients, let's say less than 10. Fortunately, that's usually not an issue, since the vast majority of fast foods have well over 10 ingredients, as we'll see shortly.

Evaluating Fast Food from McDonald's, Pizza Hut, and Taco Bell

As mentioned earlier, all of the PFI reports for the menu items of these fast food restaurants can be found online using the links below:

McDonald's: https://tinyurl.com/3xf8jyda

Pizza Hut: https://tinyurl.com/bdcmjkd3

Taco Bell: https://tinyurl.com/4fhefvxs

Furthermore, the key data has been summarized in three additional tables in Appendices B to D.

In those tables, the data is classified by menu item, caloric content, number of ingredients, PFI, and the degree of industrialization.

For a short summary of the restaurant data, the following table gives a breakdown of the degree of industrialization for each restaurant's menu. The numbers in the table represent how many food items were classified by the categories of industrialization. **Note that food items with less than 10 ingredients were not included in the counts as well as drinks from other manufacturers.**

Restaurant	Non-Industrialized	Lightly Industrialized	Moderately Industrialized	Highly Industrialized	Extremely Industrialized
McDonald's	1	0	18	51	1
Pizza Hut	0	0	1	150	0
Taco Bell	0	0	15	56	1

Key observations:

- The vast majority of food items fall in the moderately and highly industrialized categories with the latter dominating.

- Out of the nearly 300 food items evaluated, only one was determined as non-industrialized.

- Out of the nearly 300 food items evaluated, only two fell into the extremely industrialized category, indicating the most food items contained a mix of processed ingredients ranging from unprocessed to extremely processed.

- Of the three restaurants, nearly all Pizza Hut menu items were in the highly industrialized category; this result may seem surprising, but most of the Pizza Hut menu items have pizza in them or pizza-like ingredients and pizza is consistently rated as highly processed.

- The data clearly indicates that the vast majority of foods served in McDonald's, Pizza Hut, and Taco Bell can be classified as industrial-based.

Some Takes on Highly Processed Foods

Three other ways to look at a highly processed fast-food dish is to examine it in terms of (1) "food content" vs. "chemical content," (2) unprocessed and lightly processed ingredients vs. significantly processed ingredients, and (3) listing the additives in terms of their functional properties. As an example, let's look at Pizza Hut's "Tuscani Creamy Chicken Alfredo Pasta."

Here are the ingredients:

Rotini Pasta (Water(1), Enriched Semolina (Durum Wheat Semolina)(2), Niacin(3), Ferrous Sulfate(4), Thiamine Mononitrate(5), Riboflavin(6), Folic Acid(7))), Alfredo Sauce (Skim Milk(8), Water(9), Soybean Oil(10) or Canola Oil(11), Parmesan and Romano Cheeses (Part Skim Cow's Milk(12), Enzymes(13), Salt(14), Cellulose(15)), Modified Corn Starch(16), Contains 2% or Less of: Salt(17), Dried Cream (Cream(18), Nonfat Milk(19), Tocopherols(20) and Ascorbyl Palmitate(21) (To Help Protect Flavor), Butter (Cream(22), Natural Flavor(23), Salt(24), Parmesan Cheese Paste (Parmesan Cheese (Milk(25), Cheese Cultures(26), Salt(27), Enzymes(28)), Water(29), Salt(30), Sodium Phosphate(31), Sodium Citrate(32)), Garlic Puree(33), Spice(34), Cheese Flavor (Cheddar Cheese (Cultured Milk(35), Salt(36), Enzymes(37)), Water(38), Salt(39), Enzymes(40), Cultures(41), Phosphoric Acid(42), Xanthan Gum(43), Potassium Sorbate(44)), Enriched Flour (Bleached Wheat Flour(45), Niacin(46), Reduced Iron(47), Thiamine Mononitrate(48), Riboflavin(49), Folic Acid(50), DATEM(51), Mono- & Diglycerides(52), Yeast Extract(53), Xanthan Gum(54), Natural Flavors(55), Lactic Acid(56), Calcium Lactate(57), Chicken (White Meat(58), Water(59), Seasoning (Salt(60), Yeast Extract(61), Spices(62), Dried Cane Syrup(63), Dextrose(64), Carrageenan(65), Dried Chicken Broth(66), Dried Garlic(67), Dried Onion(68), Chicken Fat(69), Dried Parsley(70)), Modified Food Starch(71), Sodium Phosphates(72)), Cheese (Part

Skim Mozzarella Cheese: (Pasteurized Milk(73) and Skim Milk(74), Cheese Cultures(75), Salt(76), Enzymes(77)), Sugar Cane Fiber(78) (Added to Prevent Clumping), Modified Food Starch(79), Potassium Chloride(80), Natural Flavors(81), Rosemary Extract(82) (To Protect Flavor).

Now I'll express the ingredients in terms of foods/beverages (gray highlight) vs. chemicals (bolded type). Chemicals are defined as additives made in factories. Note that water is included in the food/beverage category.

Rotini Pasta (Food(1), Enriched Semolina (Food(2), **Chemical**(3), **Chemical**(4), **Chemical**(5), **Chemical**(6), **Chemical** (7))), Alfredo Sauce (Food (8), Food(9), Food (10) or Food (11), Parmesan and Romano Cheeses (Food (12), **Chemical** (13), Food (14), **Chemical** (15)), **Chemical** (16), Contains 2% or Less of: Food (17), Dried Cream (Food (18), Food (19), **Chemical**(20) and **Chemical** (21) (To Help Protect Flavor), Butter (Food (22), **Chemical** (23), Food (24), Parmesan Cheese Paste (Parmesan Cheese (Food (25), Food (26), Food (27), **Chemical** (28)), Food(29), Food (30), **Chemical** (31), **Chemical** (32)), Food (33), Food (34), Cheese Flavor (Cheddar Cheese (Food (35), Food (36), **Chemical** (37)), Food(38), Food (39), **Chemical** (40), Food (41), **Chemical** (42), **Chemical** (43), **Chemical** (44)), Enriched Flour (Food (45), **Chemical** (46), **Chemical** (47), **Chemical** (48), **Chemical** (49), **Chemical** (50), **Chemical** (51), **Chemical** (52), **Chemical**(53), **Chemical** (54), **Chemical** (55), **Chemical** (56), **Chemical** (57), Chicken (Food (58), Food(59), Seasoning (Food (60), **Chemical** (61), Food (62), Food (63), **Chemical** (64), Food (65), Food (66), Food (67), Food (68), Food (69), Food (70)), **Chemical** (71), **Chemical** (72)), Cheese (Part Skim Mozzarella Cheese: (Food (73) and Food (74), Food(75), Food (76), **Chemical** (77)), Food (78) (Added to Prevent Clumping), **Chemical** (79), **Chemical** (80), **Chemical** (81), **Chemical** (82) (To Protect Flavor).

In this dish, 40 out of the 82 ingredients (or 49%) are chemicals. Almost half of this menu item is chemicalized indicating how highly processed this food is.

The following list denotes the ingredients as unprocessed (**UP**) or lightly processed (**LP**) in bolded type vs. significantly processed in gray highlights:

Rotini Pasta (**UP/LP** (1), Enriched Semolina (**UP/LP**)(2), Niacin(3), Ferrous Sulfate(4), Thiamine Mononitrate(5), Riboflavin(6), Folic Acid(7))), Alfredo Sauce (**UP/LP** (8), **UP/LP** (9), Soybean Oil(10) or Canola Oil(11), Parmesan and Romano Cheeses (**UP/LP** (12), Enzymes(13), **UP/LP** (14), Cellulose(15)), Modified Corn Starch(16), Contains 2% or Less of: **UP/LP** (17), Dried Cream (**UP/LP** (18), **UP/LP** (19), Tocopherols(20) and Ascorbyl Palmitate(21) (To Help Protect Flavor), Butter (**UP/LP**(22), Natural Flavor(23), **UP/LP** (24), Parmesan Cheese Paste (Parmesan Cheese (**UP/LP** (25), **UP/LP** (26), **UP/LP** (27), Enzymes(28)), **UP/LP** (29), **UP/LP** (30), Sodium Phosphate(31), Sodium Citrate(32)), **UP/LP** (33), **UP/LP** (34), Cheese Flavor (Cheddar Cheese (Cultured Milk(35), **UP/LP** (36), Enzymes(37)), **UP/LP** (38), **UP/LP** (39), Enzymes(40), Cultures(41), Phosphoric Acid(42), Xanthan Gum(43), Potassium Sorbate(44)), Enriched Flour (Bleached Wheat Flour(45), Niacin(46), Reduced Iron(47), Thiamine Mononitrate(48), Riboflavin(49), Folic Acid(50), DATEM(51), Mono- & Diglycerides(52), Yeast Extract(53), Xanthan Gum(54), Natural Flavors(55), Lactic Acid(56), Calcium Lactate(57), Chicken (**UP/LP** (58), **UP/LP** (59), Seasoning (**UP/LP** (60), Yeast Extract(61), **UP/LP** (62), Dried Cane Syrup(63), Dextrose(64), Carrageenan(65), **UP/LP** (66), **UP/LP** (67), **UP/LP** (68), **UP/LP** (69), **UP/LP** (70)), Modified Food Starch(71), Sodium Phosphates(72)), Cheese (Part Skim Mozzarella Cheese: (**UP/LP** (73) and **UP/LP** (74), Cheese Cultures(75), **UP/LP** (76), Enzymes(77)), Sugar Cane Fiber(78) (Added to Prevent Clumping), Modified Food Starch(79), Potassium

Chloride(80), Natural Flavors(81), Rosemary Extract(82) (To Protect Flavor).

In this analysis, of the 82 ingredients, 50 (or 61%) are significantly processed confirming the earlier description of this Pizza Hut menu item.

Now let's look at the same list but from the perspective of additive types in bolded type:

Rotini Pasta (Water(1), Enriched Semolina (Food(2), **Vitamin**(3), **Mineral**(4), **Vitamin**(5), **Vitamin**(6), **Vitamin** (7))), Alfredo Sauce (Food (8), Water(9), Food (10) or Food (11) Oils, Parmesan and Romano Cheeses (Food (12), **Enzyme** (13), Food (14), **Thickener** (15)), **Texturizer** (16), Contains 2% or Less of: Food (17), Dried Cream (Food (18), Food (19), **Preservative**(20) and **Preservative** (21) (To Help Protect Flavor), Butter (Food (22), **Flavor Agent** (23), Food (24), Parmesan Cheese Paste (Parmesan Cheese (Food (25), Food (26), Food (27), **Enzyme** (28)), Water(29), Food (30), **Stabilizer** (31), **Emulsifier** (32)), Food (33), Food (34), Cheese Flavor (Cheddar Cheese (Food (35), Food (36), **Enzyme** (37)), Water(38), Food (39), **Enzyme** (40), Food (41), **Stabilizer** (42), **Thickener** (43), **Preservative** (44)), Enriched Flour (Food (45), **Vitamin** (46), **Mineral** (47), **Vitamin** (48), **Vitamin** (49), **Vitamin** (50), **Emulsifier** (51), **Emulsifier** (52), **Flavor Agent**(53), **Thickener** (54), **Flavor Agent** (55), **Acidifier** (56), **Buffer** (57), Chicken (Food (58), Water(59), Seasoning (Food (60), **Flavor Agent** (61), Food (62), Food (63), **Sweetener** (64), Food (65), Food (66), Food (67), Food (68), Food (69), Food (70)), **Texturizer** (71), **Stabilizer** (72)), Cheese (Part Skim Mozzarella Cheese: (Food (73) and Food (74), Cheese Cultures(75), Food (76), **Enzyme** (77)), Food (78) (Added to Prevent Clumping), **Texturizer** (79), **Flavor Agent** (80), **Flavor Agent** (81), **Preservative** (82) (To Protect Flavor).

There are 40 ingredients in the Pizza Hut dish that serve particular functions as additives. Think for a moment: If you were preparing this dish in your kitchen, would you use any of these additives? If not, then again, we can conclude that Pizza Hut's Tuscani Creamy Chicken Alfredo Pasta dish is highly processed.

I'm hitting you with a bunch of food additives at this point, but try not be overwhelmed. Many of these additives will be discussed in detail in Part II of the book. Also, for immediate gratification, there are two appendices in the back of the book that you can use as references to look up info on a specific additive or ingredient:

Appendix E: Food Processing Terms & Definitions

Appendix F: Glossary of Food Ingredients & Additives

Examining Ingredients in Specific Menu Items

Finally, I get to the most interesting section of the book, at least in my opinion. This is the food science section. Until now, fast food ingredients have been discussed rather indirectly, but in this section specific ingredients will be addressed and explored in some detail.

If you have checked out the Glossary of Food Ingredients & Additives in Appendix F, you found over 270 items listed. I'm not going to comment on all of those items, but I've selected about 50 of them for some detailed description. They will be scattered across a few dozen menu items taken from the three restaurants. Of course, many of the ingredients/additives appear multiple times, but I'll talk about each of them only once. Hopefully, by the end of this section of the book, you'll have a pretty good sense of how commercial foods are constructed, the functions of food additives, and the stories behind them, particularly the ones that I find interesting, surprising, shocking or all of the above.

The information about ingredients in this section was obtained from a variety of sources, including personal knowledge. I relied heavily on Ruth Winter's *A Consumer's Dictionary of Food Additives.* Several online resources were also helpful.

Further Readings:
A Consumer's Dictionary of Food Additives

https://tinyurl.com/4r5y5dyh

Food Additives Website

https://foodadditives.net/

FDA's Substances Added to Food

https://tinyurl.com/3myja3wx

Bakerpedia: Ingredients in the Baking Industry

https://bakerpedia.com/ingredients//

CHAPTER 10

KEY TAKEAWAYS IN PART I

- There are over 9000 certified food additives in the USA, most of them are synthetic, and the list grows every year.

- As part of the Nutrition Facts Label, the ingredients in a food product are listed in decreasing amounts by weight or volume except when the designation <x% is used to designate very small amounts for ingredients placed at the end of the list.

- The food industry and its supporting players, the food ingredient companies, have created synthetic foods that never existed before, but which provide exemplary texture, taste, smell, eye appeal, and shelf stability ... these are miracles of food science but are they healthy?

- New food additives fall under the classification GRAS, which stands for Generally Recognized as Safe. Although new food additives are required to be safe for consumption, the federal government, through the FDA, rarely performs investigative studies to determine their level of safety. That requirement is left to the manufacturers or their contractor labs.

- 37% of adults (~85 million) consumed fast food on any given day.

- About 11.3% of total daily calories come from fast food.

- In the years 2009 to 2014, the number of fast-food restaurants grew by 9%. In 2014, the total number was 228,677.

- Since the 1970s, fast food has become a significant part of the American diet. Between 1977 and 1978, fast food accounted for just over 3% of calories in the American diet, but, between 2005 and 2008, that amount leapt to over 13%.

- Processed foods have increasingly become physically and chemically manipulated. Food additives change color, alter taste, and perform other functions to totally change the look and appeal of food. These changes have led to reductions in food quality and potential threats to public health.

- Commercial foods are all processed in some way, but what distinguishes one from another is the degree of processing, which can range from lightly processed to highly processed.

- The Processed Food Index (PFI) provides a relative scale to compare commercial foods in terms of how industrialized they are. The PFI is not perfect, but it can be a useful tool for evaluating the "unnaturalness" of a processed food.

PART II

EXAMINING FAST FOODS

CHAPTER 11

INGREDIENTS IN SELECTED MCDONALD'S MENU ITEMS

"We do it all for you."

— McDonald's slogan

General Comments

The fast-food offerings at McDonald's cover the gamut from breakfast through dinner; as we'll see, a broad range of processed ingredients are involved in the making of their products. As I get into examining particular menu items, pay attention to (1) the number of ingredients composing the food or beverage; (2) how many highly processed ingredients are involved (refer to the PFI score for a numerical representation of that observation); and (3) how many ingredient names you don't recognize.

For each menu item addressed in the body of the text, I'll focus only on a small group of ingredients for each restaurant. These specific ingredients will be highlighted in gray in the ingredient list for each menu item. Other ingredients not mentioned in the body of the text are described in the Glossary in Appendix F.

As a reminder, all of the surveyed McDonald's menu items, including the ones mentioned below, can be viewed at: https://tinyurl.com/3xf8jyda

The McDonald's menu items selected for close-up reviews are:

- Southwest Buttermilk Crispy Chicken Salad
- Quarter Pounder® with Cheese Bacon
- Big Breakfast® with Hotcakes
- McCafé® Mocha Frappé, Medium
- McCafe'® Mango Pineapple Smoothie

McDonald's Southwest Buttermilk Crispy Chicken Salad (500 Cal)

Ingredients:

BUTTERMILK CRISPY CHICKEN FILLET: Chicken Breast Fillets with Rib Meat(1), Water(2), Vegetable Oil (Canola Oil(3), Corn Oil(4), Soybean Oil(5), Hydrogenated Soybean Oil(6)), Wheat Flour(7), Bread Crumbs (Wheat Flour(8), Vinegar(9), Sea Salt(10), Baking Soda(11), Inactive Yeast(12), Natural Flavor(13)), Potato Starch(14), Buttermilk (Cultured Nonfat Milk(15), Salt(16), Sodium Citrate(17), Vitamin A Palmitate(18), Vitamin D3(19)), Salt(20), Citric Acid(21), Rice Starch(22), Palm Oil(23), Corn Starch(24), Rice Flour(25), Corn Flour (Yellow) (26), Natural Flavors(27), Spices(28), Baking Soda(29), Sugar(30), Garlic Powder(31), Xanthan Gum(32), Maltodextrin(33).

SALAD BLEND: Romaine Lettuce(34), Baby Spinach(35), Carrots(36), Baby Kale(37), Lollo Rossa Lettuce(38), Red Leaf Lettuce(39), Red Oak Lettuce(40), Red Tango Lettuce(41), Red Romaine Lettuce(42), Red Butter Lettuce(43).

SOUTHWEST VEGETABLE BLEND: Roasted Corn(44), Black Beans(45), Roasted Tomato(46), Poblano Pepper(47), Lime Juice (Water(48), Lime Juice Concentrate(49), Lime Oil(50)), Cilantro(51).

CILANTRO LIME GLAZE: Water(52), Corn Syrup Solids(53), High Fructose Corn Syrup(54), Sugar(55), Distilled Vinegar(56), Olive Oil(57), Soybean Oil(58), Freeze-Dried Orange Juice Concentrate(59), Cilantro(60), Xanthan Gum(61), Preservatives (Sodium Benzoate(62) and Potassium Sorbate(63)), Garlic Powder(64), Spice(65), Propylene Glycol Alginate(66), Onion Powder(67), Citric Acid(68).

SHREDDED CHEDDAR/JACK CHEESE: Cheddar Cheese (Pasteurized Milk(69), Cheese Culture(70), Salt(71), Enzymes(72), Annatto(73) [Color]), Monterey Jack Cheese (Pasteurized Milk(74), Cheese Culture(75), Salt(76), Enzymes(77)), Potato Starch(78), Corn Starch(79), Dextrose(80), Powdered Cellulose(81) (Prevent Caking), Calcium Sulfate(82), Natamycin (83) (Natural Mold Inhibitor), Enzyme(84).

CHILI LIME TORTILLA STRIPS: Corn(85), Vegetable Oil (Corn(86), Soybean(87) and Sunflower Oil(88)), Salt(89), Maltodextrin(90), Sugar(91), Dried Tomato(92), Dextrose(93), Spices(94), Onion Powder(95), Green Bell Pepper Powder(96), Citric Acid(97), Autolyzed Yeast Extract(98), Malic Acid(99), Paprika Extract(100) (Color), Disodium Inosinate(101), Disodium Guanylate(102), Natural Flavor(103), Lemon Extract(104), Spice Extractive(105).

Degree of Processing

Annatto (73) Moderately; **Autolyzed Yeast Extract** (98) Highly; **Baking Soda** (synthetic) (11,29) Extremely; **Calcium Sulfate** (synthetic) (82) Extremely; **Canola Oil** (3) Highly; **Citric Acid** (synthetic) (21,68,97) Extremely; **Corn Flour** (26) Moderately; **Corn Oil** (4) Highly; **Corn Starch** (24,79) Highly; **Corn Syrup Solids** (synthetic) (53) Extremely; **Dextrose** (synthetic) (80,93) Extremely; **Disodium Guanylate** (102) Extremely; **Disodium Inosinate** (101) Extremely; **Enzymes** (72,77,84) Highly; **High Fructose Corn Syrup** (synthetic) (54) Extremely; **Hydrogenated Soybean Oil**

(synthetic) (6) Extremely; **Malic Acid** (synthetic) (99) Extremely; **Maltodextrin** (33,90) Highly; **Natamycin** (83) Highly; **Natural Flavors** (13, 27,103) Moderately; **Paprika Extract** (100) Moderately; **Potassium Sorbate** (synthetic) (63) Extremely; **Potato Starch** (14,78)Moderately; **Powdered Cellulose** (81) Highly; **Propylene Glycol Alginate** (synthetic) (66) Extremely; **Rice Flour** (25) Moderately; **Rice Starch** (22) Highly; **Sodium Benzoate** (synthetic) (62) Extremely; **Sodium Citrate** (synthetic) (17) Extremely; **Soybean Oil** (5,58) Highly; **Spice Extractive** (105) Highly; **Sugar** (30,91) Moderately; **Sunflower Oil** (88) Highly; **Vinegar** (9,56) Moderately; **Vitamin A Palmitate** (synthetic) (18) Extremely; **Vitamin D3** (19) Extremely; **Wheat Flour** (7,8) Moderately; **Xanthan Gum** (synthetic) (32) Extremely.

Reporting the Processed Food Index (PFI) for the Crispy Chicken Salad

# OF UNPROCESSED INGREDIENTS:	18
# OF LIGHTLY PROCESSED INGREDIENTS:	18
# OF MODERATELY PROCESSED INGREDIENTS:	9
# OF HIGHLY PROCESSED INGREDIENTS:	12
# OF EXTREMELY PROCESSED INGREDIENTS:	17

TOTAL # OF INGREDIENTS:	105
# OF UNIQUE INGREDIENTS:	74 (70%)
PROCESSED FOOD INDEX:	34 (Moderately Industrialized)

Commentary

The first thing that strikes me about this menu item is that it's a salad but one with incredible complexity. It has 105 ingredients with 74 of them unique! Just imagine how long it would take you to make this salad at home from scratch. Your kitchen counters would be covered with various bags, boxes, and other containers. It would take you hours to construct this meal. A menu item with a high number of ingredients is the first clue that you've encountered a highly processed food. Surprisingly, the Processed Food Index (PFI) is only 34, representing a moderately industrialized food item, so this dish sits in the middle of the industrialization scale. The numerous vegetables, spices, and herbs in the salad counterbalance the more processed ingredients.

[Chemist's Note: *It's very important to understand that every additive that winds up in a commercial food is there for a purpose. It serves a particular function. Most food manufacturers pay particularly close attention to what goes into their products to minimize cost, so every ingredient they use must have a practical purpose. Keep in mind that behind every commercial food there is a ton of research and testing carried out prior to market launch to assure a good product at the lowest possible cost. Also, I will be making frequent reference to the FDA (Federal Food & Drug Administration), an agency which approves all ingredients and additives in commercial foods. Subsequently, I will just refer to the acronym, FDA.*]

The highlighted ingredients in McDonald's Southwest Buttermilk Crispy Chicken Salad are:

- Annatto

- Autolyzed Yeast Extract

- Disodium Guanylate & Disodium Inosinate

- Malic Acid

- Natamycin

Annatto:

This material is a vegetable dye found in cheese. It's a trickster ingredient. The original Cheddar cheese, made hundreds of years ago in England, was characteristically light orange due to the types of grasses dined on by the dairy cows. The orange color of the cheese got associated with the quality of the cheese, so cheesemakers in America, in order to mimic the English product, started to add dyes during Cheddar cheese production. Also, the buttermilk in natural milk contributes a yellow color. When milk is de-fatted, it becomes whiter, so reduced-fat milk produces white cheddar cheese. Annatto, a natural product, which comes from the seeds of the achiote tree, native to tropical areas of Mexico and Brazil, adds the orange tint that people expect in Cheddar cheese. As a food additive, annatto is generally safe, but a small part of the population is allergic to it.

Autolyzed Yeast Extract:

Found in the Chili Lime Tortilla Strips, this additive is a flavor agent made from yeast. The process for making this additive originated in the 19th century. When yeast flakes are immersed in salt water, the cells shrivel and open up releasing their contents into the water (note: autolyzed means self-destroying). When the mixture is filtered to remove the solid components, a flavorful liquid remains (note: an extract is a mixture of components removed from a food by dispersion in a suitable medium or solvent). The yeast extract can be concentrated to form a paste or completely dried to produce a powder. This additive is generally safe, but, people sensitive to monosodium glutamate (MSG), should avoid it.

Disodium Guanylate/Disodium Inosinate:

These two additives, also found in the Chili Lime Tortilla Chips, are like Siamese twins since they are always found together in processed foods. As flavor enhancers, they have little taste of their own, but they accentuate the natural flavors of foods, particularly in combination. They can boost bland flavors and transform a sad-tasting food into a great one. Although these chemicals are found naturally in many foods, as additives in commercial foods, they are most likely obtained from fermentation processes. Japanese food scientists were the first to investigate these types of flavor enhancers. In 1912, Shintaro Kodama discovered disodium guanylate in fish flakes. Decades later, Akira Kuninake isolated disodium inosinate from yeast extract using an enzymatic process. The additives are generally considered safe, but they do get converted to uric acid in the body, so people with gout should avoid them.

[**Chemist's Note:** Food additives are most useful when they can be dissolved in water. Insoluble chemicals can often be made soluble by chemical conversion to their sodium counterparts. However, these additives contribute another source of sodium in the diet, so people with hypertension (high blood pressure) need to be aware of them. The sodium content on a food label will not only reflect the presence of salt in the food, but also any other ingredient that contains sodium.]

Malic Acid:

This additive is also used in the Chili Lime Tortilla Strips. In chemical terms, malic acid is an example of a weak acid, an acid that's not corrosive to the skin. There are a number of food additives that are in the category of weak acids. In food science, these substances are called acidulants. Just like the name sounds, they add acidity to a food product or, another way of putting it, they lower the pH. The acidity can help in preserving the food and adds a sour taste to the product. Malic acid is naturally found in many fruits, e.g., apples; in fact, the name comes from malum, the Latin word for apple from which it was first isolated. Also, biologically, all living organisms can make it. Although this additive has natural sources, the food industry uses a synthetic version. Malic acid is considered a safe additive.

Natamycin:

When you say this word, it might sound familiar. Have you heard the word streptomycin, a medical antibiotic? So too is natamycin (also called pimaricin). Why is an antibiotic used as a food additive? You find natamycin listed with the cheese ingredients. It's sourced from a Strep bacterium found in soil. It serves as a fungicide to inhibit the formation of mold on the cheese, so it's a type of preservative. However, it's not a bactericide. As it's considered safe for humans, the FDA allows its use in foods. In medicine, natamycin is an antifungal agent used to treat eye infections. In the food industry, it's used in dairy and meat products.

[Chemist's Note: Preservatives are some of the most common food additives. Food manufacturers want their products to last as long as possible on store shelves. This is called shelf life. Spoilage of processed foods affects a company's bottom line and is avoided at all costs. There are basically two kinds of preservatives: (1) antioxidants and (2) antimicrobials. The former remove oxygen which can causes fats and oils to go rancid. The latter inhibits the formation of bacteria, mold, and yeast.]

McDonald's Quarter Pounder® with Cheese Bacon (620 Calories)

Ingredients:

BUN: Enriched Flour (Wheat Flour(1), Malted Barley Flour(2), Niacin(3), Reduced Iron(4), Thiamine Mononitrate(5), Riboflavin(6), Folic Acid(7)), Water(8), Sugar(9), Yeast(10), Soybean Oil(11), Contains 2% or Less: Sesame Seeds(12), Salt(13), Wheat Gluten(14), Dextrose(15), Guar Gum(16), Vinegar(17), Vegetable Proteins (Pea(18), Potato(19), Rice(20)), Sunflower(21) and/or Canola

Oil(22), Maltodextrin(23), Natural Flavors(24), May Contain One or More Dough Conditioners (DATEM(25), Ascorbic Acid(26), Mono and Diglycerides(27), Enzymes(28)), Modified Food Starch(29).

BEEF: 100% Pure USDA Inspected Beef(30).

PASTEURIZED PROCESS AMERICAN CHEESE: Milk(31), Cream(32), Water(33), Sodium Citrate(34), Salt(35), Cheese Cultures(36), Citric Acid(37), Enzymes(38), Soy Lecithin(39), Color Added(40)*.

PICKLE SLICES: Cucumbers(41), Water(42), Distilled Vinegar(43), Salt(44), Calcium Chloride(45), Alum(46), Potassium Sorbate (Preservative)(47), Natural Flavors(48), Polysorbate 80(49), Extractives of Turmeric (Color)(50).

ONIONS: Chopped Onion(51).

MUSTARD: Distilled Vinegar(52), Water(53), Mustard Seed(54), Salt(55), Turmeric(56), Paprika(57), Spice Extractive(58).

KETCHUP: Tomato Concentrate from Red Ripe Tomatoes(59), Distilled Vinegar(60), High Fructose Corn Syrup(61), Corn Syrup(62), Water(63), Salt(64), Natural Flavors(65).

GRILL SEASONING: Salt(66), Pepper(67).

THICK CUT APPLEWOOD SMOKED BACON: Pork Bellies(68) Cured with Water(69), Salt(70), Sugar(71), Natural Smoke Flavor(72), Sodium Phosphate(73), Sodium Erythorbate(74), Sodium Nitrite(75).

* "Color Added" is a generic phrase which could include any number of colored substances. The FDA allows this kind of nonspecific wording in food labels. Note that this ingredient, since it cannot be identified, is not classified according to its Degree of Processing.

Ingredient Assignments and the Degree of Industrialization (Processing)

Alum (synthetic) (46) Extremely; **Ascorbic Acid** (synthetic) (26) Extremely; **Calcium Chloride** (synthetic) (45) Extremely; **Canola Oil** (22) Highly; **Citric Acid** (37) Highly; **Corn Syrup** (62) Highly; **DATEM** (synthetic) (25) Extremely; **Dextrose** (Synthetic) (15) Extremely; **Enzymes** (28,38) Highly; **Extractives of Turmeric** (50) Highly; **Folic Acid** (synthetic) (7) Extremely; **Guar Gum** (16) Moderately; **High Fructose Corn Syrup** (synthetic) (61) Extremely; **Malted Barley Flour** (2) Moderately; **Maltodextrin** (23) Highly; **Modified Food Starch** (29) Extremely; **Mono and Diglycerides** (synthetic) (27) Extremely; **Natural Flavors** (24, 48, 65) Moderately; **Natural Smoke Flavor** (72) Highly; **Niacin** (synthetic) (3) Extremely; **Pea Protein** (18) Moderately; **Polysorbate 80** (synthetic) (49) Extremely; **Potassium Sorbate** (synthetic) (47) Extremely; **Potato Protein** (19) Highly; **Reduced Iron** (synthetic) (4) Moderately; **Riboflavin** (synthetic) (6) Extremely; **Rice Protein** (20) Highly; **Sodium Citrate** (synthetic) (34) Extremely; **Sodium Erythorbate** (synthetic) (74) Extremely; **Sodium Nitrite** (synthetic) (75) Extremely; **Sodium Phosphate** (synthetic) (73) Extremely; **Soy Lecithin** (39) Highly; **Soybean Oil** (11) Highly; **Spice Extractives** (58) Highly; **Sugar** (9,71) Moderately; **Sunflower Oil** (21) Highly; **Thiamine Mononitrate** (synthetic) (5) Extremely; **Vinegar** (17,43,52,60) Moderately; **Wheat Flour** (1) Moderately; **Wheat Gluten** (14) Moderately.

Reporting the Processed Food Index (PFI) for the Quarter Pounder® with Cheese

# OF UNPROCESSED INGREDIENTS:	11
# OF LIGHTLY PROCESSED INGREDIENTS:	5
# OF MODERATELY PROCESSED INGREDIENTS:	9
# OF HIGHLY PROCESSED INGREDIENTS:	13
# OF EXTREMELY PROCESSED INGREDIENTS:	18
TOTAL # OF INGREDIENTS:	75

 # OF UNIQUE INGREDIENTS: 56 (75%)

PROCESSED FOOD INDEX: 47 (Highly Industrialized)

Commentary

Of immediate note is the fact that there are 75 ingredients in this item with 56 (75%) of them being unique. That's sounds like a bunch of ingredients for a hamburger, but, before alarms go off, note that there are a number of processed products used in the making of this sandwich including the bun, processed American cheese, pickles, mustard, ketchup, and bacon. If you made a bacon cheeseburger at home using store-bought products, you would wind up with a similar number of ingredients. Here are some store-bought items with the number of ingredients in parentheses:

- Ball Park Hamburger Buns (26)

- Heinz Tomato Ketchup (8)

- Heinz Hamburger Dill Chips (12)

- Heinz Tomato Ketchup (8)

- Kraft American Processed Cheese (15)

- Oscar Mayer Bacon (7)

Just counting the above hamburger fixings gives 76 ingredients. So, in terms of ingredients, there is nothing extraordinary about the number of ingredients found in this McDonald's hamburger. It's just surprising to realize how many ingredients wind up in a seemingly simple food like a bacon cheeseburger.

As regards calories, notice that this single food item accounts for about 26% of the recommended daily intake for the average American.

The highlighted ingredients in McDonald's Quarter Pounder® with Cheese Bacon are:

- Ascorbic Acid

- DATEM

- Maltodextrin

- Natural Flavor

- Polysorbate 80

Ascorbic Acid:

Ascorbic acid has multiple functions as a food additive. In this case, it's used in preparing the bun for the hamburger. It functions, along with a few other ingredients, as a dough conditioner. The dough is improved by strengthening the gluten (which holds it together), providing a larger loaf volume, increasing the tenderness of the bread, helping with rising, and making the bread appear fresher.

In another application, ascorbic acid is a preservative and acts as an antioxidant in commercial foods. In processed foods, oxygen from the air can oxidize fatty or oily substances creating off-flavors and unpleasant odors. Ascorbic acid binds the oxygen and prevents the negative effects of oxidation. A common name for this additive is Vitamin C, but food manufacturers usually don't list it that way because they are not adding it as a nutrient. A positive trend in the food industry in recent years is to replace potentially harmful preservatives, such as butylated hydroxytoluene (BHT) and butylated hydroxyanisole (BHA), with harmless preservatives like ascorbic acid.

DATEM:

This additive sounds like a word, but it's really an acronym. Sometimes on food labels you will even see it written as Datem, as if it was a word and not an acronym. The FDA allows .the use of a handful of acronyms on food labels. I'm not sure why, but it may be to save space because the names are often very long. Unfortunately, acronyms are not useful to most consumers who likely won't recognize what they stand for. DATEM represents Diacetyl Tartaric Acid Esters of Mono- and Diglycerides. Quite a mouthful, even for a chemist! DATEM is actually a mixture of compounds, but I'll refer to it as a single compound. Its primary function is to act as an emulsifier, a substance which binds together oil and water. Ordinarily, oil and water won't mix, so, over time, oil will rise to the surface of a food, which is not an appealing look to a consumer. Emulsifiers act like the eggs in mayonnaise to keep oil and water from separating. DATEM can also serve as a dough conditioner to improve bread volume and uniformity. That's its purpose in the hamburger, where it shows up in the bun. It can make up 0.375 to 0.5% of the weight of flour used in commercial baking.

To explain how this synthetic additive is made would require the reader to have a degree in chemistry, so I won't go there. Suffice it to say that the process is complicated, and that's why the additive is tagged as extremely processed. The FDA recognizes DATEM as safe, but excessive consumption of the additive may reduce the bioavailability of nutrients, particularly calcium, and some medical researchers claim that it contributes to leaky gut syndrome. Note that the Whole Foods Company has banned the sale of products containing DATEM. Some people think that gluten-sensitive people may actually be reacting to other chemicals, like DATEM, in bread products and not to the gluten itself. DATEM is now a fixture in the modern baking industry, but it's a chemical that your great grandparents probably weren't exposed to.

Maltodextrin:

This additive sounds like it's a specific substance, but, wait, you're being fooled. It's really a generic term for a group of substances which can be sourced from a variety of grains like corn, rice, and wheat. As mentioned earlier, when corn starch is completely broken down, glucose forms. If corn starch is only partially broken down, it can form several groups of water-soluble chemicals collectively called maltodextrins. The less the starch is broken down, the more the maltodextrin acts like starch. Conversely, the greater the breakdown, the more the maltodextrin acts like glucose. So, there is a range of maltodextrins from not very sweet to pretty sweet. You'll never see a specific group listed on a food label ... just the generic term "maltodextrin." To make things even more complex, maltodextrins can be prepared from any source of starch, e.g., wheat, rice, potato, corn, etc. and usually you won't see that on the label either.

In this case, the maltodextrin is used to make the hamburger bun. It provides some bulk for the dough, some sweetness, may help to stabilize the bun, and increases the shelf life. Although very industrial in nature, maltodextrins are safe ingredients to consume.

Natural Flavor(s):

If you are thinking that a natural flavor is the opposite of an artificial flavor, you would be only partly correct. The word "natural" carries all kinds of positive associations, like improved health, and food companies liberally use that term to hook consumers. But that word can be deceptive, as we'll see.

On commercial food labels, natural flavor is the fourth most common ingredient. Food manufacturers want their foods to taste good, so consumers will become repeat customers, but they also want their product to stand out among their competitors. The phrase "natural flavor" is generic. There are hundreds, if not thousands, of natural flavor ingredients. Companies protect them as trade secrets (proprietary formulas), and the FDA allows them to hide their identities on food labels. You may see the phrase in the singular or the plural. Note that a single flavor is usually a mixture of compounds. People with allergies to individual flavoring agents are out of luck since they will not see them listed on the ingredient list.

According to the FDA, a natural flavor is defined as follows: "A flavor ingredient that has components derived from plant, meat, seafood, dairy, or fermentation products." The source is natural but the final product not so much. Various scientific techniques are used to isolate and concentrate flavor components oftentimes involving toxic organic solvents. Residues from the toxic solvents may wind up as contaminants in the flavoring ingredient.

As an example, McDonald's uses a beef flavor to add a unique taste to their French fries. Flavor chemists, having identified the amino acids released in the cooking of beef, construct a flavor ingredient with the same amino acids plus other flavor enhancing compounds to create a meaty "natural" beef flavor. When McDonald's switched from using beef tallow to 100% vegetable oil to fry the potatoes, they added a beef flavor in the par-frying of the potatoes which kept the fries tasting the same as before. They listed "beef flavor" as "natural flavor," but that ingredient disqualified the fries as a vegetarian food. McDonald's got sued in 2001 for that misidentification. Today, the "beef" flavor is derived from hydrolyzed wheat and milk proteins and probably qualifies as vegetarian, but McDonald's does not certify it as such.

Polysorbate 80:

Polysorbate 80 (also called Tween 80) definitely sounds like it comes from a chemical lab, and it does. Polysorbates are amber-colored, viscous liquids that are ubiquitous in the biopharmaceutical and food industries. They serve as emulsifiers, which, as mentioned earlier, keep oily components dispersed in water. They keep fast-food shakes intact, maintain the creaminess of salad dressings, and hold ice cream components together. In the case of the McDonald's hamburger, the polysorbate 80 is found in the pickles, where it's not acting as an emulsifier but more like a preservative. Polysorbates come in different "numerical flavors" such as polysorbate 20, 40, 60, and 80. The higher the number the greater the attraction to the oily substances in the food. As additives, polysorbates can be used up to 0.1% of the volume of food.

The synthesis of polysorbates is a complex process, and a knowledge of organic chemistry would be essential to understanding how it's made. It all starts with sorbitol, a sugar alcohol, which is chemically dehydrated (water removed) and reacted with ethylene oxide (toxic gas) and a long-chain fatty acid, like oleic acid, to form polymeric molecules. The result is not a single substance, but a mixture of chemically similar molecules.

Polysorbates have been implicated in several health issues. Studies with rats have shown that polysorbates can initiate low-grade inflammation, weight gain, metabolic problems, and disrupt the bacterial composition of the gut microbiome. They may worsen Crohn's disease.

A very recent application of polysorbate 80 is its use in the preparation of Johnson & Johnson's COVID-19 vaccine during the 2020 pandemic.

McDonald's Big Breakfast® with Hotcakes (1340 Calories)

Ingredients:

BISCUIT: Enriched Flour (Bleached Wheat Flour(1), Niacin(2), Reduced Iron(3), Thiamine Mononitrate(4), Riboflavin(5), Folic Acid(6)), Cultured Nonfat Buttermilk (Cultured Skim Milk(7), Nonfat Dry Milk(8), Modified Food Starch(9), Salt(10), Mono and Diglycerides(11), Locust Bean Gum(12), Carrageenan(13)), Palm Oil(14), Palm Kernel Oil(15), Water(16), Leavening (Baking Soda (Sodium Bicarbonate)(17), Sodium Aluminum Phosphate(18), Monocalcium Phosphate(19)), Contains 2% or Less: Salt(20), Sugar(21), Modified Cellulose(22), Wheat Protein Isolate(23), Natural Flavor(24), Modified Food Starch(25), Xanthan Gum(26), Soy Lecithin(27).

HASH BROWNS: Potatoes(28), Vegetable Oil (Canola Oil(29), Soybean Oil(30), Hydrogenated Soybean Oil(31)), Natural Beef Flavor(32) [Wheat and Milk Derivatives], Salt(33), Corn Flour(34), Dehydrated Potato(35), Dextrose(36), Sodium Acid

Pyrophosphate(37) (maintain color), Extractives of Black Pepper(38).

HOTCAKES: Water(39), Enriched Flour (Wheat Flour(40), Niacin(41), Reduced Iron(42), Thiamine Mononitrate(43), Riboflavin(44), Folic Acid(45)), Whey(46), Corn Flour(47), Soybean Oil(48), Sugar(49), Eggs(50), Leavening (Baking Soda (Sodium Bicarbonate)(51), Sodium Aluminum Phosphate(52), Monocalcium Phosphate(53)), Dextrose(54), Emulsifier (Mono and Diglycerides(55), Propylene Glycol Monoesters of Fats and Fatty Acids(56), Sodium Stearoyl Lactylate(57)), Salt(58), Xanthan Gum(59), Natural Flavors(60), Beta Carotene(61) (Color), Soy Lecithin(62).

HOTCAKE SYRUP: Corn Syrup (63), Sugar (64), Water(65), Natural Flavors(66), Potassium Sorbate(67) (Preservative), Caramel Color(68).

SCRAMBLED EGGS: Eggs (69), Citric Acid (70).

SAUSAGE PATTY: Pork (71), Water (72), Salt(73), Spices(74), Dextrose(75), Sugar(76), Rosemary Extract(77), Natural Flavors(78).

SALTED BUTTER: Cream (79), Salt(80).

CLARIFIED BUTTER: Pasteurized Cream (Butterfat(81)).

SALTED WHIPPED BUTTER: Sweet Cream(82), Salt(83).

Ingredient Assignments and the Degree of Industrialization (Processing)

Baking Soda (synthetic) (17,51) Extremely; **Beta Carotene** (61) Highly; **Bleached Wheat Flour** (1) Highly; **Canola Oil** (29) Highly; **Caramel Color** (68) Highly; **Carrageenan** (13) Moderately; **Citric Acid** (synthetic) (70) Extremely; **Corn Flour** (34) Moderately; **Corn Syrup** (63) Highly; **Dextrose** (synthetic) (36,54) Extremely; **Extractives of Black Pepper** (38) Highly; **Folic Acid** (synthetic) (6,45) Extremely; **Hydrogenated Soybean Oil** (31) Extremely; **Locust Bean Gum** (12) Highly; **Modified Cellulose** (22) Highly;

Modified Food Starch (9,25) Extremely; **Mono and Diglycerides** (synthetic) (11,55); Extremely; **Monocalcium Phosphate** (synthetic) (19,53); Extremely; **Natural Beef Flavor** (32) Highly; **Natural Flavors** (24,60) Moderately; **Niacin** (synthetic) (2,41) Extremely; **Palm Kernel Oil** (15) Moderately; **Potassium Sorbate** (Synthetic) (67) Extremely; **Propylene Glycol Monoesters of Fats and Fatty Acids** (Synthetic) (56) Extremely; **Reduced Iron** (3,42) Moderately; **Riboflavin** (synthetic) (5,44) Extremely; **Rosemary Extract** (77) Highly; **Sodium Acid Pyrophosphate** (synthetic) (37) Extremely; **Sodium Aluminum Phosphate** (synthetic) (18,52) Extremely; **Sodium Stearoyl Lactylate** (synthetic) (57) Extremely; **Soy Lecithin** (27) Highly; **Soybean Oil** (30,48) Highly; **Sugar** (21,49,64) Moderately; **Thiamine Mononitrate** (synthetic) (4,43) Extremely; **Wheat Flour** (40) Moderately; **Wheat Protein Isolate** (23) Moderately; **Xanthan Gum** (synthetic) (26); Extremely.

Reporting the Processed Food Index (PFI) for the Big Breakfast® with Hotcakes:

# OF UNPROCESSED INGREDIENTS:	5
# OF LIGHTLY PROCESSED INGREDIENTS:	10
# OF MODERATELY PROCESSED INGREDIENTS:	8
# OF HIGHLY PROCESSED INGREDIENTS:	12
# OF EXTREMELY PROCESSED INGREDIENTS:	17
TOTAL # OF INGREDIENTS:	83
# OF UNIQUE INGREDIENTS:	52 (63%)
PROCESSED FOOD INDEX:	48 (Highly Industrialized)

Commentary

The first thing I notice is the energy content of this breakfast (1340 calories). Given the average daily intake of 2000 calories, this breakfast alone provides 67% of the caloric needs of the day. That doesn't leave much for lunch, dinner, and snacks. A reasonable explanation for the high caloric value of this meal is the presence of 11 ingredients that are sources of fat and oil, such as vegetable oils, dairy products, and meat. A good fact to remember is fat and oil have more than double the calories of carbohydrates and protein per gram.

There are 52 unique ingredients because there's a lot happening in this menu item: biscuit, hash browns, hotcakes, syrup, scrambled eggs, sausage patty, and three types of butter. Of the 52 unique ingredients, 37 of them, or 71%, are classified as significantly processed, which explains the PFI score of 48 (highly industrialized food).

The highlighted ingredients in McDonald's Big Breakfast® with Hotcakes are:

- Dextrose
- Extractives of Black Pepper
- Hydrogenated Soybean Oil
- Modified Food Starch
- Mono- and Diglycerides

Dextrose:

In chemistry, this basic sweetener is called dextrose; in medicine, it's called glucose (blood sugar). The terms are interchangeable. As a food ingredient, it has been used for many, many years. Although it delivers a sweet note to food, this simple sugar is only 0.75 times as sweet as table sugar (sucrose). In McDonald's breakfast meal, dextrose is used in both the hash browns and the hotcakes. Only in America would potatoes need to be sweetened!

Although the primary function of dextrose is to sweeten food products, it also may be used as a humectant (attracts and retains moisture) or to create a soft mouthfeel (pleasant physical sensation in mouth). In rare cases, it can be used as a preservative since bacteria cannot survive in high concentrations of sugars.

[**Chemist's Note**: *The reason you don't have to refrigerate maple syrup or honey is because of the very high sugar content. Those foods will not spoil at room temperature. The scientific name for that phenomenon is osmotic pressure in which bacterial cells collapse due to the huge difference in sugar concentration between the outside and inside of the bacterial cells. To correct the imbalance, water will leach out of the bacterial cells causing them to shrivel and die.]*

Most dextrose is manufactured from starch, which can be obtained from various types of plants. In the USA, corn starch is the primary source of dextrose. The production of corn starch was described earlier in the discussion about high fructose corn syrup. Crystalline dextrose, a solid form, is obtained by removing the water from dextrose solutions.

Extractives of Black Pepper:

This is a specific example of a group of flavor additives called spice extractives. There are two types of spice extractives: essential oils and oleoresins. The latter are used in foods. The "oleo" in oleoresins refers to oil, so an oleoresin is a mixture of natural oils and resins obtained from plants. In general, resins are liquid compounds which can be induced to harden by exposure to light or a chemical reaction. Two well-known examples are amber and balsam. A common, non-food example of an oleoresin is turpentine.

In the case of spices, like black pepper, a physical process called solvent extraction is used. The raw spice is cleaned, dried, and ground and then exposed to a volatile, organic solvent to extract the oleoresin. Then the solvent is stripped off by some physical means to leave only the oleoresin. The complete removal of the solvent is essential since in some cases the solvents used are toxic. In general, oleoresins are 5 to 20 times stronger in flavor than the dried spices. They are more potent in flavor than their essential oil cousins. For commercial use, the extracts are dispersed in a dry carrier or suitable liquid, e.g., vegetable oil, to some desired concentration for use by a food manufacturer. Since they have properties of the original spice, an extractive is easy to store and transport because concentrates need less space. They are more heat stable and provide a longer shelf life than the raw spices. Some food applications for spice extractives are:

- As a coloring agent in butter, cheese, meats, snack foods, and cereals

- As a flavor agent for various foods

- In jellies, jams, and gelatin preparations

- In frozen foods, desserts, soups, fish preserved in oil, meat sauces, or any prepared food where a more vibrant color is desired

In the McDonald's hotcake meal, the extractive of black pepper shows up in the hash browns as a flavoring agent. Spice extractives are common ingredients in processed foods. Other examples include extractives of paprika, celery, rosemary, and turmeric.

Hydrogenated Soybean Oil:

If you've seen a bottle of soybean oil sitting on a grocery store shelf, you may recall that the contents are a light-yellow liquid. Of course, soybean oil comes from soybeans. Soybeans, compared to other legumes, are rich in oil with a content of about 18% and are about double that in protein. It is the most widely used oil in the world. In the soybean industry, the soybeans are crushed, some moisture is removed, the residue is heated and then rolled into flakes. A gasoline-type solvent mix called hexanes is combined with the flakes to extract the oil in a solvent-extraction process. The solvent is then stripped off for re-use. The crude oil is then refined to remove unwanted impurities.

To increase the shelf-life of soybean oil, it's put through a process called hydrogenation. In that process, the oil is mixed with hydrogen gas and a metal catalyst at elevated temperatures which converts unsaturated components to saturated ones. The final product is a solid fat, i.e., hydrogenated soybean oil, not subject to rancidity. This process is similar to, but not the same, for the production of partially hydrogenated fats, which are currently banned in commercial foods due to the formation of trans fats, chemicals linked to the occurrence of heart disease.

In the McDonald's hotcakes meal, the hydrogenated soybean oil is used in the hash browns as part of a mixture with canola and soybean oils.

Modified Food Starch:

When you see the phrase "modified food starch" note that it's a generic term. There are dozens of types of starch derived from plants, e.g., corn starch, wheat starch, rice starch, etc., so this ingredient is not very useful as far as information goes. The food manufacturer really doesn't want you to know exactly what they're using in their product.

A very common food starch is corn starch, so let's look at that example. However, modified corn starch is still a generic term. Confused yet? The word "modified" implies that the corn starch has been physically, chemically or enzymatically altered in some way. The fact that the food manufacturer is not required to reveal the type of modified corn starch in the cereal product is not helpful to the consumer who has no idea what they're eating. There are dozens of chemicals that can react with corn starch to modify it. Some of these chemicals are very hazardous, and the FDA specifies acceptable and safe residue levels for these chemicals.

Modifying the original starch is done for a variety of reasons such as stabilizing the starch against temperature changes, serving as a thickening agent or emulsifier, making the starch more dissolvable, and many other purposes. Modified food starches are usually synthesized in a manufacturing plant using such hazardous chemicals as propylene oxide, acetic anhydride, sulfuric acid, bleach, etc. The FDA approves each modified starch as safe for human consumption, but modified corn starches, unlike their plant-based starting material, are anything but natural. Here are some examples of modified food starches that are lumped under that generic title: bleached starch, acetylated starch, hydroxypropylated starch, and sodium octenyl succinate starch. That innocent little phrase, "modified corn starch", really represents a plethora of industrial modifications, mostly chemical.

In the McDonald's hotcakes meal, modified food starch shows up twice in the biscuits. Of course, we don't know what the starch basis is or what type of modification was used, but according to the Bakerpedia website, in general, modified <u>wheat</u> starches act as thickeners, provide superior freeze-thaw stability (the food holds up with temperature changes), and offer help with moisture control among other benefits.

Mono- and Diglycerides

First, a brief, simplified chemistry lesson. When any fat, plant or animal, is broken down into its constituent parts you get glycerol, a type of alcohol, and three fatty acid molecules, which could be all the same or different. So, a typical fat (or oil) is called a <u>triglyceride</u>. Chemists have figured out ways to make these types of molecules with only one or two fatty acid molecules, thus creating mono- and diglycerides.

These additives can act as dough conditioners, emulsifiers (agents for mixing water and oil), or stabilizers. Additionally, they can increase aeration in a batter and improve the mixing of dry and liquid ingredients. This is a very common food additive, particularly in baked goods. It is estimated that every American consumes about half-pound of these chemicals every year. Since mono- and diglycerides are found naturally in normal fats, they are considered harmless as an additive; however, as additives in commercial foods they are synthetic substances manufactured in chemical plants.

In the McDonald's hotcakes meal, mono- and diglycerides are ingredients both in the biscuit and the hotcakes.

McDonald's McCafé® Mocha Frappé, Medium (510 Calories)

Ingredients:

MOCHA COFFEE FRAPPE BASE: Cream(1), Skim Milk(2), Sugar(3), Water(4), High Fructose Corn Syrup(5), Coffee Extract(6), Milk(7), Natural(8) & Artificial(9) Flavors, Cocoa(10) (Processed with Alkali), Mono & Diglycerides(11), Guar Gum(12), Potassium Citrate(13), Disodium Phosphate(14), Carrageenan(15), Locust Bean Gum(16), Red 40(17), Yellow 5(18), Blue 1(19).

WHIPPED LIGHT CREAM: Cream(20), Nonfat Milk(21), Liquid Sugar(22), Contains 2% or Less: Mono and Diglycerides(23), Natural Flavors(24), Carrageenan(25).

CHOCOLATE DRIZZLE: Corn Syrup(26), Dextrose(27), Water(28), Sugar(29), Glycerin(30), Hydrogenated Coconut Oil(31), Cocoa(32) (Processed with Alkali), Modified Food Starch(33), Nonfat Milk(34), Natural Flavors(35), Salt(36), Gellan Gum(37), Disodium Phosphate(38), Potassium Sorbate(39) (Preservative), Soy Lecithin(40).

ICE: Ice(41).

Ingredient Assignments and the Degree of Industrialization (Processing)

Artificial Flavors (synthetic) (9) Extremely; **Blue 1** (19) Extremely; **Carrageenan** (15,25) Moderately; **Coffee Extract** (6) Moderately; **Corn Syrup** (synthetic) (26) Highly; **Dextrose** (synthetic) (27)

Extremely; **Disodium Phosphate** (synthetic) (14,38) Extremely; **Glycerin** (synthetic) (30) Extremely; **Gellan Gum** (37) Moderately; **Guar Gum** (12) Moderately; **High Fructose Corn Syrup** (synthetic) (5) Extremely; **Hydrogenated Coconut Oil** (synthetic) (31) Extremely; **Liquid Sugar** (22) Moderately; **Locust Bean Gum** (16) Highly; **Modified Food Starch** (33) Extremely; **Mono and Diglycerides** (synthetic) (11,23) Extremely; **Natural Flavors** (8,24) Moderately; **Potassium Citrate** (synthetic) (13) Extremely; **Potassium Sorbate** (synthetic) (39) Extremely; **Red 40** (synthetic) (17) Extremely; **Soy Lecithin** (40) Highly; **Sugar** (3,29) Moderately; **Yellow 5** (synthetic) (18) Extremely.

Reporting the Processed Food Index (PFI) for the Mocha Frappe

# OF UNPROCESSED INGREDIENTS:	2
# OF LIGHTLY PROCESSED INGREDIENTS:	5
# OF MODERATELY PROCESSED INGREDIENTS:	7
# OF HIGHLY PROCESSED INGREDIENTS:	3
# OF EXTREMELY PROCESSED INGREDIENTS:	13

TOTAL # OF INGREDIENTS:	41
# OF UNIQUE INGREDIENTS:	30 (73%)
PROCESSED FOOD INDEX:	53 (Highly Industrialized)

Commentary

Of immediate note is the fact that there are 41 ingredients in this item and 30 of them are unique. That's a bunch of ingredients for just one drink. Also of note is the observation that 43% of the unique ingredients are classified as extremely processed, which accounts for the high PFI value of 53.

As regards caloric load, notice that this beverage has 510 calories, accounting for about 26% of the recommended daily intake for the average American.

We live in the era of specialty beverages. Most every fast-food restaurant offers its own proprietary concoctions. Think for a moment about the enormous task of creating these unique beverages. It certainly takes months, if not years, to bring these new products to the marketplace. On the technical side, a team of chemists, food scientists, technicians, flavorists, taste testers, and other professionals need to formulate, trial and evaluate the new product to make sure the taste, appearance, and texture are just right for consumers. Coming up with the complex combination of 41 ingredients for the mocha frappe is no simple matter. On the business side, the best and/or cheapest suppliers of the ingredients must be located and contracts arranged. Marketing and advertising plans need to be designed and implemented. Possibly, new machines and serving containers may need to be designed, manufactured, and placed in the stores. Introducing new beverages and foods into a fast-food lineup is incredibly expensive costing in the millions of dollars.

The highlighted ingredients in McDonald's McCafé ® Mocha Frappé are:

- Artificial Flavor
- FD&C Food Colors (Red 40, Yellow 5, Blue 1)
- High Fructose Corn Syrup

Artificial Flavor:

This is not the name of an ingredient. It is, however, the generic name of a group of ingredients. In some cases, the FDA allows food manufacturers to use generic names instead of specific ingredients. Most likely, the reason is to protect proprietary formulas, or trade secrets, so other companies won't steal their creations.

Unfortunately, for consumers, this lack of transparency could unknowingly cause health issues for people who may be sensitive or allergic to the chemicals used. Of course, the word "artificial" alerts you to the fact that the flavor is synthetically made in a factory. Why do food companies use artificial flavors? First, taste is critical for a food product, so getting the taste just right is essential. Second, like the huge variety of perfumes and fragrances, these flavors can be tailor-made to mimic most any natural flavor. Also, they are likely less expensive than their natural counterparts.

Some of the largest companies in the world are flavor houses whose chemists (flavorists) create thousands of tasty chemicals. Since they primarily supply commercial food companies and don't market to the general public, you're likely unfamiliar with their names. Three of the biggest companies are: Firmenich (Swiss), Givaudan (Swiss), and Sensient (USA). Sensient has almost 4000 employees worldwide and earned about 1.5 billion dollars in 2013. How many flavors does it offer? From the website: "All of them. Literally. We created an infinite line of world-class flavor options to help our partners' products achieve brand definition, greater appeal, increased patent compliance and more." Some of the flavors from these companies mimic real foods, like "roast chicken type flavor" or "grilled" flavor or "fruity" flavor. Some of the artificial flavors don't have a taste of their own, but they'll enhance flavors like sweetness or savoriness.

FD&C Food Colors (Red 40, Yellow 5, and Blue 1):

Red 40, Yellow 5, and Blue 1 are artificial dyes and examples of food colorings used to convey a desired color to a food product. FD&C stands for Food, Drug, and Cosmetic, a designation that the FDA has approved these chemicals as colorants in food. You'll often see the FD&C designation on food labels, e.g., FD&C Red 40. Red 40 is also known as Allura Red. It imparts a red color to foods and beverages. Yellow 5 is also known as tartrazine and imparts a yellow-orange color to foods and beverages. Blue 1 is also known as Brilliant Blue and imparts a blue color to foods and beverages. Why are all three colors in the Mocha Coffee Frappe? Red, yellow, and blue are primary colors. Mixing them together gives some shade of black. In combination with the other ingredients, the final color is probably brownish. Note that there is no actual coffee in this product, so artificial dyes are used to mimic the expected color of coffee. In the food industry, food dyes have a colorful history, pun intended. They were originally synthesized from chemicals found in coal tar in the late 1800's, a by-product of the coal industry, when coal was carbonized to make coke. Sometimes they are called "coal tar dyes," but not these days ... too unappealing! Coal tar is a thick black complex liquid. It has had a variety of industrial uses including paving roads, sealant for asphalt, as a medicine in shampoos, soaps, and ointments, and as a topical medicine. However, on the downside, coal tar contains carcinogenic substances. From 1850 to 1900 chemists figured out how to extract chemicals from coal tar that could be converted to highly-colored dyes. They were looking for cheap, easily made dyes to replace natural ones. Natural dyes were hard to come by, often expensive, and availability was unpredictable. If artificial dyes could be created in a laboratory and scaled up in a factory, food companies would no longer need natural ones, and they could save a bunch of money. Plus, the artificial dyes were stable, mass producible, and had very long shelf lives.

In 1906, there were 700 food dyes available in the USA. Because of health issues, the US Pure Food and Drug Act of 1906 reduced the permitted list of artificial dyes to 7. With improvements in chemical analysis and purification techniques in the 20th century, the list of approved food dyes once again began to grow. After WWII, food companies started to use massive amounts of artificial dyes in processed foods. However, by the 1950's scientists were discovering that many food dyes caused health problems. The FDA started looking into the issue and began banning some of the artificial dyes used in the food industry. For example, Orange 1, found to cause liver cancer, was banned in 1966. Yellow 1 and Yellow 2 were banned in 1960 due to causing intestinal lesions. Over the decades, numerous food dyes have been delisted and removed from the marketplace, so by 2016, there were only 7 dyes still approved for use by the food industry. Notice that the approved artificial food colors that we see today tend to have high numbers, since their predecessors have been banned or replaced. For example, Red 40 had 39 predecessors! Still, today, there are some concerns about the remaining FD&C dyes. Red 40 is suspected to cause problems for hyperactive children. Aspirin-sensitive people may react to the consumption of Yellow 5. Some people may be allergic to Blue 1.

It's interesting to note that regulatory agencies around the world do not agree with one another regarding the safety of food dyes. For example, in Europe the dyes can be used in foods, but the products are required to carry a warning label which says "may have an adverse effect on activity and attention in children." There are no such label requirements in the USA. Slowly, very slowly, food companies are replacing hazardous, artificial colors with safer ones sourced from plants, e.g., carotene, found in carrots, imparts a yellow-orange color to foods.

High Fructose Corn Syrup:

The use of high fructose corn syrup (HFCS) as a sweetener in processed foods has become ubiquitous. Any Generation X person or younger takes it for granted that HFCS is part of the food supply. But, HFCS is a fairly young sweetener. Let's get into a little bit of history and chemistry. [Don't worry, I'll keep it as simple as I can!] First, let's talk sugar (sucrose) for which the main agricultural sources are sugar beets and sugar cane. Sugar can be chemically broken down into glucose and fructose. It's interesting to note that fructose alone is about 1.7 times sweeter than sucrose, so if you added crystalline fructose to your tea, it would taste sweeter than if you added the same amount of table sugar.

Prior to the 1960s, Cuba was a big supplier of cane sugar to the USA, but with the overthrow of the government by Fidel Castro and the subsequent cutting of trade ties, the USA had to rely on the sugar supplies from other countries. As a result, the price of sugar not only went up, but it fluctuated with the world market. The instability of the sugar supply in the USA prompted research into alternative sweeteners. Commercial corn syrup, made from corn starch, had been around since the early 1900s, but traditional corn syrup is not as sweet as cane sugar, so it could not be used as a substitute in foods and beverages. In 1957, two chemists discovered an enzyme that could convert glucose into fructose. However, it took about 15 years to develop the technology to carry out the conversion on an industrial scale, to carefully mimic the sweetness of cane sugar, and to match or come under the price of cane sugar. Then food manufacturers, particularly beverage companies, had to be convinced to switch over. The availability of HFCS had a number of advantages over cane sugar: (1) There was an abundance of corn in the USA at stable prices; (2) The production of corn was subsidized by the government to keep prices low and the availability high; (3) Since HFCS was a liquid, it was easily transportable by truck and rail car; and (4) Also, as a liquid, HFCS could easily be stored in manufacturing facilities in vats and tanks, then pumped from one area of the plant to another.

I classify HFCS as an extremely processed ingredient. The following production process will explain why: (1) Shucked corn is delivered to the factory; (2) The corn is steeped in a slightly acidic medium to soften it up and then coarsely ground to allow for the separation of protein (gluten), fat (germ), and fiber in a mechanical process. What remains is a slurry of corn starch; (3) This slurry is treated with a mixture of enzymes and acid at an elevated temperature to break down the starch into glucose; (4) The mixture is filtered to remove impurities and then the liquid portion is decolorized; (5) The glucose mixture is exposed to an enzyme (glucose isomerase) to convert glucose to fructose. The resulting liquid sweetener has only 42% fructose, which is not as sweet as table sugar; (6) Next, the 42% HFCS is put through a liquid chromatographic process (enzymes embedded on a column) to increase the fructose content to 55% HFCS, equivalent to the sweetness of table sugar. The name "high fructose corn syrup" comes from the enrichment of the glucose syrup with fructose. The aforementioned complex, multi-step process makes HFCS an extremely processed food ingredient.

McD's McCafe´® Mango Pineapple Smoothie (250 Cal, Medium)

Ingredients:

MANGO PINEAPPLE FRUIT BASE: Water(1), Clarified Demineralized Pineapple Juice Concentrate(2), Mango Puree Concentrate(3), Pineapple Juice Concentrate(4), Orange Juice Concentrate(5), Pineapple Puree (Pineapple(6) and Ascorbic Acid(7)), Apple Juice Concentrate(8), Contains Less Than 1%: Cellulose Powder(9), Natural(10) and Artificial Flavors(11), Xanthan Gum(12), Pectin(13), Citric Acid(14), Fruit(15) and Vegetable Juice(16) for Color, Turmeric Extract(17) (Color).

LOWFAT SMOOTHIE YOGURT: Cultured Grade A Reduced Fat Milk(18), Sugar(19), Whey Protein Concentrate(20), Fructose(21), Corn Starch(22), Modified Food Starch(23), Gelatin(24), Active Yogurt Cultures(25). **ICE:** Ice(26).

Ingredient Assignments and the Degree of Industrialization (Processing)

Artificial Flavors (synthetic) (11) Extremely; **Ascorbic Acid** (synthetic) (7) Extremely; **Cellulose Powder** (9) Moderately; **Citric**

Acid (synthetic) (14) Extremely; **Corn Starch** (22) Highly; **Fructose** (synthetic) (21) Extremely; **Gelatin** (24) Extremely; **Modified Food Starch** (23) Extremely; **Natural Flavors** (10) Moderately; **Pectin** (13) Highly; **Sugar** (19) Moderately; **Turmeric Extract** (17) Moderately; **Whey Protein Concentrate** (20) Moderately; **Xanthan Gum** (synthetic) (12) Extremely.

Reporting the Processed Food Index (PFI) for the Mango Pineapple Smoothie

# OF UNPROCESSED INGREDIENTS:	4
# OF LIGHTLY PROCESSED INGREDIENTS:	8
# OF MODERATELY PROCESSED INGREDIENTS:	5
# OF HIGHLY PROCESSED INGREDIENTS:	2
# OF EXTREMELY PROCESSED INGREDIENTS:	7

TOTAL # OF INGREDIENTS:	26
# OF UNIQUE INGREDIENTS:	26 (100%)
PROCESSED FOOD INDEX:	35 (Moderately Industrialized)

Commentary

This is the second McCafe´® beverage to be reviewed. The first one was the Mocha Frappé. I'm going to place the data side-by-side to more easily see the differences.

Beverage	# Ingredients	PFI	# Calories	# Extremely Processed Ingredients	# Added Sugar Ingredients
Mocha Frappé	41	53	510	13	6
Mango	26	35	250	7	2

Pineapple
Smoothie

If you're at all concerned about health, it's easy to see which one would be the better choice. Hands down, the smoothie would be better for you: fewer ingredients, much lower PFI, over half the calorie content, about half the number of extremely processed ingredients, and one-third as many added sugar sources. Of course, I'm not naïve enough to think that people go into fast food restaurants to select the healthiest food items, but, with this data in hand, a health-conscious consumer could make much better choices.

The highlighted ingredients in McDonald's McCafé ® Mango Pineapple Smoothie are:

- Gelatin

- Xanthan Gum

Gelatin:

Found in one of the most beloved desserts, Jell-O, gelatin is one of many thickening or gelling agents found in processed foods. They are designed to add texture to foods and beverages, and, particularly in the latter, they serve to increase mouthfeel. What would you rather drink, a thin, vapid beverage or a thick, creamy one?

Gelatin, first manufactured in the mid-nineteenth century, is a translucent, colorless, flavorless powder derived from animal hooves and joints, particularly from pigs and cows. The making of gelatin is a complex process. First, animal parts are washed, then soaked in hot water to remove grease. Next, the parts are roasted for 30 minutes at 200°F, after which they are soaked in acid or alkali for about 5 days to remove minerals. Again, the animal parts are boiled in water this time to dissolve the collagen. Then the solution of collagen is sterilized by flash-heating at 375°F, followed by filtration to remove residue, and then cooled to precipitate the gelatin. Finally, machines separate the liquid and solid fractions (gelatin), the solids are dried, and the final product is pulverized to produce a powder. The complexity of this process is the reason why gelatin, a natural product, was tagged as extremely processed.

Xanthan Gum:

Xanthan gum is a plant-based thickener which provides viscosity (thickness) and stabilization to many foods and beverages. Typically, the additive is derived from the fermentation of corn-derived dextrose by bacterial organisms (*Xanthomonas campestris*) that have been genetically modified to maximize the excretion of the gum as a product of their metabolism. A complicated industrial process turns the viscous material into a cream-colored powder. By the way, the same bacterium is responsible for the black rot found on cauliflower and broccoli plants.

What About a Whole McDonald's Meal?

All of the food and beverage examinations so far have dealt with individual menu items. Let's take a quick look at a classic McDonald's meal: Big Mac®, Fries, and Chocolate Shake.

BIG MAC®:

BUN: Enriched Flour (Wheat Flour(1), Malted Barley Flour(2), Niacin(3), Iron(4), Thiamine Mononitrate(5), Riboflavin(6), Folic Acid(7)), Water(8), Sugar(9), Yeast(10), Soybean Oil(11), Contains 2% or Less: Salt(12), Wheat Gluten(13), Potato Flour (14), May Contain One or More Dough Conditioners (DATEM(15), Ascorbic Acid(16), Mono and Diglycerides(17), Enzymes(18)), Vinegar(19).

BEEF: 100% Beef Patty(20).

LETTUCE: Shredded lettuce (21).

BIG MAC SAUCE: Soybean Oil(22), Sweet Relish (Diced Pickles(23), Sugar(24), High Fructose Corn Syrup(25), Distilled Vinegar(26), Salt(27), Corn Syrup(28), Xanthan Gum(29), Calcium Chloride(30), Spice Extractives(31)), Water(32), Egg Yolks(33), Distilled Vinegar(34), Spices(35), Onion Powder(36), Salt(37), Propylene Glycol Alginate(38), Garlic Powder(39), Vegetable Protein (Hydrolyzed Corn, Soy, and Wheat(40)), Sugar(41), Caramel

Color(42), Turmeric(43), Extractives of Paprika(44), Soy Lecithin(45).

PASTUERIZED PROCESS AMERICAN CHEESE: Milk(46), Cream(47), Water(48), Sodium Citrate(49), Salt(50), Cheese Cultures(51), Citric Acid(52), Enzymes(53), Soy Lecithin(54), Color Added(55).

PICKLE SLICES: Cucumbers(56), Water(57), Distilled Vinegar(58), Salt(59), Calcium Chloride(60), Alum(61), Potassium Sorbate (Preservative)(62), Natural Flavors(63), Polysorbate 80(64), Extractives of Turmeric (Color)(65). **ONIONS:** Chopped Onion(66).

GRILL SEASONING: Salt(67), Pepper(68).

FRENCH FRIES: Potatoes(69), Vegetable Oil (Canola Oil(70), Corn Oil(71), Soybean Oil(72), Hydrogenated Soybean Oil(73), Natural Beef Flavor(74) [wheat And Milk Derivatives]), Dextrose(75), Sodium Acid Pyrophosphate(76) (maintains Color), Salt(77).

CHOCOLATE SHAKE:

VANILLA REDUCED FAT ICE CREAM: Milk(78), Sugar(79), Cream(80), Corn Syrup(81), Natural Flavor(82), Mono and Diglycerides(83), Cellulose Gum(84), Guar Gum(85), Carrageenan(86), Vitamin A Palmitate(87).

CHOCOLATE SHAKE SYRUP: Sugar(88), Corn Syrup(89), Water(90), Glycerin(91), Cocoa (92) (Processed with Alkali), Fruit and Vegetable Juice(93) (Color), Salt(94), Natural Flavors(95), Potassium Sorbate(96) (Preservative).

WHIPPED LIGHT CREAM: Cream(97), Nonfat Milk(98), Liquid Sugar(99), Contains 2% or Less: Mono and Diglycerides(100), Natural Flavors(101), Carrageenan(102).

This meal has a total of 102 ingredients, not all unique of course, but just think about all the time, technology, and research that went into the selection of the ingredients for each portion of this meal. The gray highlighted ingredients represent those that I label significantly processed. There are 67 of them. As a percentage, they account for 66% of the total. Not unexpectedly, this meal is very processed and that should give one pause when consuming it. The consequences of diets high in very processed foods will be discussed later in Part III.

What about the caloric content of this meal? The Big Mac® delivers 540 calories; the World-Famous French Fries contribute 320 calories; and the Chocolate Shake provides an incredible 630 calories, even more than the hamburger. The total calories are 1490 or 75% of the daily recommendation for an average person. Eat like that every day, and you're likely to rapidly expand your waistline.

CHAPTER 12

INGREDIENTS IN SELECTED PIZZA HUT MENU ITEMS

"Better Ingredients. Better Pizza"

— Pizza Hut slogan

General Comments

As expected, the majority of the menu items for Pizza Hut involve pizza in one form or another. The pizzas are composed of a crust, a sauce, cheese(s), and various toppings of meats and vegetables. The Pizza Hut menu is huge because there are myriad combinations of those 5 components creating many dozens of possibilities, most of which are tagged with unique brand names. For example, a pizza, with a combination of crust, sauce, cheese, and meat is not just called a meat pizza, but it gets the flashier name of Meat Lover's® Pizza.

Here is a list of the various pizza crusts with their acronyms.

- Hand Tossed Crust (HTC)

- Original Stuffed Crust (OSC)

- Original Pan Crust (OPC)

- Personal Pan Crust (PPC)

- Rectangular Crust (RC)

- Thin 'N Crispy Crust (TNC)

• Udi's® Gluten-Free Crust

The rest of the Pizza Hut menu is filled out by pasta dishes, salads, sides, and desserts.

Of the three fast-food restaurants examined in this book, the online ingredient lists at the Pizza Hut website are the least comprehensive. There are dozens of items whose compositions are not reported, particularly "local specialties", regional menu items, soups and sandwiches.

As a reminder, all of the surveyed Pizza Hut menu items, including the ones mentioned below, can be viewed at: https://tinyurl.com/bdcmjkd3

The Pizza Hut menu items selected for close-up looks are:

• **Crispy Chicken Caesar Salad**

• **Tuscani® Creamy Chicken Alfredo Pasta**

• **Super Supreme Personal Pan Pizza® Slice**

• **Meat Lover's® Original Stuffed Crust® Slice**

• **The Great Beyond Original Pan Pizza® Slice**

• **Cinnabon® Stick**

Pizza Hut's Crispy Chicken Caesar Salad (830 Calories)

Ingredients:

SEASONED CROUTONS: Enriched Flour (Wheat Flour(1), Malted Barley Flour(2), Niacin(3), Reduced Iron(4), Thiamine Mononitrate(5), Riboflavin(6), Folic Acid(7)), Canola(8) and/or Sunflower Oil(9), Whey(10), Salt(11), Yeast(12), 2% or Less of: High Fructose Corn Syrup(13), Sugar(14), Onion Powder(15), Dehydrated Parsley(16), Calcium Propionate(17) (Preservative), Calcium Peroxide(18), Calcium Sulfate(19), Ascorbic Acid(20),

Azodicarbonamide(21), Enzymes(22), Sodium Stearoyl Lactylate(23), Spice Extractive(24), Spices(25), Paprika(26) (Color), Turmeric(27) (Color), Extractive of Paprika(28) (Color), Citric Acid(29), TBHQ(30) (To Preserve Freshness).

PARMESAN CHEESE: Pasteurized Milk(31), Cheese Cultures(32), Salt(33), Enzymes(34)), Powdered Cellulose(35) As An Anticaking Agent.

LEAFY GREENS BLEND: Iceberg Lettuce(36), Green Leaf Lettuce(37).

GARLIC PARMESAN MEDIUM BONE-OUT WING: Bone-Out Uncooked Chicken Breast Chunk Fritters with Rib Meat(38) Containing Up To 20% Of A Solution Of: Water(39), Seasoning (Salt(40), Corn Maltodextrin(41), Potassium Chloride(42), Garlic Powder(43), Onion Powder(44), Chicken Fat(45), Flavor(46) [Including Extractives of Celery(47)], Chicken Broth(48), Autolyzed Yeast Extract(49)), Modified Food Starch(50), Potassium(51) and Sodium Phosphates(52), Battered With: Water(53), Wheat Flour(54), Salt(55), Leavening (Sodium Acid Pyrophosphate(56), Sodium Bicarbonate(57), Monocalcium Phosphate(58)), Spice(59), Onion Powder(60)), Breaded With: Wheat Flour(61), Salt(62), Leavening (Sodium Acid Pyrophosphate(63), Sodium Bicarbonate(64)), Spices(65), Onion Powder(66), Pre-Dusted With: Wheat Flour(67), Wheat Gluten(68), Salt(69), Leavening (Sodium Acid Pyrophosphate(70), Sodium Bicarbonate(71)), Chicken Fat(72), Spice(73), Chicken Broth(74), Onion Powder(75), Flavor(76), Breading Set in Vegetable Oil, Fry Oil: Soybean Oil(77), Hydrogenated Soybean Oil(78) with TBHQ(79) and Citric Acid(80) added as preservatives, and Dimethylpolysiloxane(81) added as an anti-foaming agent.

SAUCE: Soybean Oil(82), Water(83), Parmesan Cheese (Pasteurized Milk(84), Cheese Culture(85), Salt(86), Enzymes(87)), Salt(88), Sugar(89), Distilled Vinegar(90), Contains Less Than 2% Of: Dehydrated Garlic(91), Enzyme Modified Egg Yolk (Egg Yolk(92), Salt(93), Phospholipase Enzyme(94), Polysorbate 60(95), Phosphoric Acid(96), Lactic Acid(97), Sodium Benzoate(98) and Potassium Sorbate(99) (As Preservatives), Spices(100), Xanthan Gum(101), Oleoresin Carrot(102) and Paprika(103) (Color), Calcium Disodium EDTA(104) Added to Protect Flavor, Natural Flavors(105), Sprinkled With Parmesan Parsley Mix: Hard Grating Cheese (Cultured Part-Skim Milk(106), Salt(107), Enzymes(108), Romano Cheese From Cow's Milk (Cultured Part-Skim Milk(109), Salt(110), Enzymes(111), Powdered Cellulose(112) (Anti-Caking Agent), Potassium Sorbate(113) (Preservative), Dried Parsley Flakes(114).

Ingredient Assignments and the Degree of Industrialization (Processing)

Ascorbic Acid (synthetic) (20) Extremely; **Autolyzed Yeast Extract** (49) Highly; **Azodicarbonamide** (synthetic) (21) Extremely; **Calcium Disodium EDTA** (synthetic) (104) Extremely; **Calcium Peroxide** (synthetic) (18) Extremely; **Calcium Propionate** (synthetic) (17) Extremely; **Calcium Sulfate** (synthetic) (19) Extremely; **Canola Oil** (8) Highly; **Citric Acid** (synthetic) (29,80) Extremely; **Corn Maltodextrin** (41) Highly; **Dimethylpolysiloxane** (synthetic) (81) Extremely; **Distilled Vinegar** (90) Moderately; **Enzymes** (22,34,87,108,111) Highly; **Extractives of Celery** (47) Highly; **Extractive of Paprika** (28) Highly; **Flavor** (Synthetic) (46,76) Extremely; **Folic Acid** (synthetic) (7) Extremely; **High Fructose Corn Syrup** (synthetic) (13) Extremely; **Hydrogenated Soybean Oil** (synthetic) (78) Extremely; **Lactic Acid** (synthetic) (97) Extremely; **Malted Barley Flour** (2) Moderately; **Modified Food Starch** (50) Extremely; **Monocalcium**

Phosphate (synthetic) (58) Extremely; **Natural Flavors** (105) Moderately; **Niacin** (synthetic) (3) Extremely; **Oleoresin Carrot** (102) Highly; **Oleoresin Paprika** (103) Highly; **Phospholipid Enzyme** (94) Highly; **Phosphoric Acid** (synthetic) (96) Extremely; **Polysorbate 60** (synthetic) (95) Extremely; **Potassium Chloride** (synthetic) (42) Extremely; **Potassium Phosphate** (synthetic) (51) Extremely; **Potassium Sorbate** (synthetic) (99,113) Extremely; **Powdered Cellulose** (35,112) Highly; **Reduced Iron** (4) Moderately; **Riboflavin** (synthetic) (6) Extremely; **Sodium Acid Pyrophosphate** (synthetic) (56,63,70) Extremely; **Sodium Benzoate** (synthetic) (98) Extremely; **Sodium Bicarbonate** (synthetic) (57,64,71) Extremely; **Sodium Phosphates** (synthetic) (52) Extremely; **Sodium Stearoyl Lactylate** (synthetic) (23) Extremely; **Soybean Oil** (77,82) Highly; **Spice Extractive** (24) Highly; **Sugar** (14,89) Moderately; **Sunflower Oil** (9) Highly; **TBHQ** (synthetic) (30,79) Extremely; **Thiamine Mononitrate** (synthetic) (5) Extremely; **Wheat Flour** (1,61,67) Moderately; **Wheat Gluten** (68) Moderately; **Whey** (10) Moderately; **Xanthan Gum** (synthetic) (101) Extremely.

Reporting the Processed Food Index (PFI) for Crispy Chicken Caesar Salad

# OF UNPROCESSED INGREDIENTS:	5
# OF LIGHTLY PROCESSED INGREDIENTS:	17
# OF MODERATELY PROCESSED INGREDIENTS:	8
# OF HIGHLY PROCESSED INGREDIENTS:	13
# OF EXTREMELY PROCESSED INGREDIENTS:	30

TOTAL # OF INGREDIENTS:	114
# OF UNIQUE INGREDIENTS:	73 (64%)
PROCESSED FOOD INDEX:	52 (Highly Industrialized)

Commentary

The first observation for this menu item is the incredible number of ingredients. Just reading the list takes your breath away. With 114 ingredients, this is a complex salad indeed. A typical salad that I make in my kitchen has maybe 20 ingredients and most of those come from the dressing. As far as I can tell, the ingredient list above does not include dressing! This Pizza Hut salad has 73 unique ingredients, so there are 73 separate foods and additives that are combined to create this dish.

The second observation is the caloric content of the salad. At 830 calories, this dish wouldn't be a good selection to support a weight-loss program. Eating this salad with a dressing would garner you half the recommended calories for the day irrespective of any other food or beverages consumed for the day.

The third observation is that most of the processed ingredients addressed below come from the croutons, which have a total of **30** constituents. The organic croutons I use on my salad from Fresh Gourmet have only 12 ingredients and are lightly processed.

The highlighted ingredients in Pizza Hut's Crispy Chicken Caesar Salad are:

- Azodicarbonamide
- Calcium Peroxide
- Dimethylpolysiloxane
- Sodium Stearoyl Lactylate
- Thiamine Mononitrate
- Wheat Flour

Azodicarbonamide (ADA):

This synthetic chemical is used in the manufacturing of flour and shows up in the croutons. It functions as a whitening agent, i.e., a bleach, and as a dough conditioner (improves the properties of dough). Azodicarbonamide is considered GRAS by the FDA and is allowed as an additive in flour up to 45 part-per-million (ppm). Interestingly, the same chemical is used as a blowing agent to make foam products, such as yoga mats. Note that many chemicals found in fast foods have uses in other industries, but that doesn't make them automatically suspect as a health hazard. However, ADA does have a controversial history.

Both Australia and the European Union countries have phased out the use of ADA. A breakdown product called semicarbazide, formed at high temperatures, such as in the bread making process, is a suspected carcinogen based on rodent trials. The World Health Organization (WHO) has linked the consumption of ADA to asthma. In the United States, due to bad publicity concerning the chemical, some fast-food restaurants have stopped using it, e.g., Subway, McDonald's, Chick-fil-A, Wendy's, White Castle, and Jack in the Box.

Further Readings:
Chicago Tribune, *The Yoga-Mat's Quiet Fast-food Exit*
https://tinyurl.com/yc76pyhp
Wikipedia, *Azodicarbonamide*
https://en.wikipedia.org/wiki/Azodicarbonamide

Calcium Peroxide:

This synthetic chemical is a rather unusual substance to appear on an ingredient list for a salad. It's present in the salad croutons. You may be familiar with the word "peroxide" as a chemical agent which removes color as in "peroxide blonde," which used to refer to a woman who turned her hair whitish blonde using peroxide chemicals. Likewise, calcium peroxide, a crystalline substance, has a whitening effect on wheat flour and is also a conditioning agent, which strengthens dough for better texture and moisture retention. The FDA allows its use up to 75ppm in flour, and it appears to be safe at that level. The chemical has other applications in the food industry as a bleaching agent for oils and a starch modifier. In other industries, it's used in soil remediation, oral hygiene, and to promote seed germination.

Dimethylpolysiloxane (PDMS):

This name is a mouthful for a non-chemist. This synthetic chemical, made from silicone oil, is also called dimethicone or Antifoam A. The latter name indicates its purpose. It serves as an anti-foaming agent for the oil used to make the breaded chicken in the salad. Foam formation can be a major problem in industrial food preparation, so this material diminishes or prevents that from happening and limits oil splatter. Dimethylpolysiloxane is also used as a lubricant and preservative. The FDA considers this additive as non-toxic. However, to preserve PDMS, formaldehyde may be used at up to 1% of the level of PDMS. Formaldehyde is a well-known preservative, but it is toxic and a human carcinogen. PDMS has many other applications including the manufacturing of Silly Putty, caulks, silicone grease and other lubricants, mold release agents, and cosmetics.

Sodium Stearoyl Lactylate (SSL):

Here's another additive with a complex chemical name found in the croutons. Similar to the azodicarbonamide and the calcium peroxide, it is added to improve the properties of the wheat dough used to make the croutons. It strengthens the gluten in the dough, so it is less likely to break or get sticky during manufacturing. Also, the dough will rise better giving a larger loaf volume. In general, SSL makes the dough more robust and better able to take a beating during processing. SSL finds wide-spread use in the baking industry, desserts, powdered beverages, cream liquors, dips, sauces, gravies, chewing gum, and pet food.

Levels in baked goods range from 0.25 to 0.50% depending on the type of flour with a typical usage at 0.375%. The additive is well tolerated by most people and is considered safe, although some consumers may have a food intolerance for products containing SSL, which could lead to digestive issues, chronic symptoms, and illness.

Thiamine Mononitrate:

Vitamins frequently appear in processed foods. Thiamine Mononitrate is a form of Vitamin B1 (another form that may appear is Thiamine Monochloride). Why are micronutrients added to foods? The history in the United States of nutrient fortification in foods goes back many years. The first example in 1924 was the addition of iodine to salt (iodized salt) to prevent goiters. Later, in the 1940s, Vitamin D was added to milk. In the late 1930s, the link between pellagra (a life-threatening disease symptomized by inflamed skin, diarrhea, and dementia) and a deficiency of Vitamin B3 (niacin) was established. During WWII, the US military observed nutritional deficiencies in new recruits, so the government established a standard of identity for wheat flour which required the addition of vitamins … this was called "enriched" flour. Since bread was a food staple in the country, the enrichment of it helped to stave off nutritional deficiencies and potential health disorders. For example, a deficiency in thiamine can cause encephalopathy (brain disease) and beriberi (symptoms are difficulty walking, paralysis of lower legs, mental confusion, speech problems, pain, and vomiting). The federal government did not require that the baked goods industry enrich flour, but the industry cooperated with the recommendations of the government, since the bakeries could not sell their baked goods to the military or other government organizations unless the products contained enriched flour.

Other micronutrients that show up in enriched flour include riboflavin (Vitamin B2), folic acid (Vitamin B9), and iron. [Note that folic acid was added in 1998 to prevent neural tube defects in newborns.] Essentially, all of the B-vitamins and iron that appear in the ingredient lists of processed and fast foods are present only due to the enrichment of wheat flour. See the next entry for more information about the nutrient quality of wheat flour.

Wheat Flour:

Wheat flour is so ubiquitous in the US food supply, whether in fast food or other types of processed foods, that one would think that, as a food ingredient, wheat flour would be innocuous and not subject to criticism. However, I have labeled it as "moderately" processed ... not as bad as highly or extremely processed ingredients, but still a bit of a concern. Note that when the phrase "wheat flour" shows up on a food label, it invariably refers to white flour or bleached flour and not whole wheat flour. Let's dive into a bit of history regarding wheat flour and the distinctions between coarse and dark flour (whole wheat) and its more popular but less healthy derivative, white flour.

For most of recorded human history, breads were made from flours using coarse grinds of wheat, barley, rye, and others. The breads were dark, not white, but they did have a tendency to acquire mold and fungus giving rise to several diseases and, due to the rancidity of fat in the flour, the shelf-life was short. Over time people developed a preference for white bread, which held up better over time, was more refined (not coarse), and produced a desirable soft texture. But white wheat flour was expensive to produce! Here's why. What makes regular wheat flour (or, as we call it today, whole wheat) coarse and dark are the parts of the wheat berry (the seed) known as the germ (the oily part) and the bran (the fibrous part). To remove those parts requires time, labor, and machinery. What's left after those parts are removed is the endosperm, the starchy part.

After grinding or milling the endosperm, it's a matter of sifting or bolting (another name for sifting) to remove the gritty particles. Even the Romans sifted wheat into 7 grades of flour using cloth made from woven horse hair. As large particles got sifted out, the flour became finer and whiter. In the old days, it took 3 tradesmen to make white flour: the miller (grinder), the boulter (sifter), and the baker. Of course, this refined flour was more expensive than whole-grain flour, so only the upper classes could afford it. [Note: Today the exact reverse is true.] The lower classes, the poor and relatively poor, prized the "high quality" bread that the wealthy ate. As early as the 13th century in England, King Henry III attempted to standardize bread making, so people would know what they were getting when they paid for bread. The law was called the Assize of Bread and Ale and was enforced until the early 1800s. According to the book, *The Dark History of Food Fraud* by Bee Wilson, the Assize regulated the weight, price, and composition of bread. It listed 7 different kinds of loaves. The best of the best was "wastel bread": it was very white and eaten by the rich. Next came "cocket bread", similar to wastel but made with an inferior flour. "Simnel bread" was a rich, cake-like food, also eaten by the rich. The stuff eaten by the masses was whole wheat bread. There were several other categories for the really poor. White bread was so prized that many bakers doctored cheap flour so it would mimic the more expensive white flour. A mineral called alum, composed of aluminum sulfate and other salts, was added to dark flour. Aluminum is not natural to the body, and alum has properties of being astringent (shrinks tissue) and emetic (causes vomiting), so the unethical bakers had to be pretty careful in using it. However, the alum would turn second-rate white flour (gray color) into a light, white, porous loaf, which could be sold at a lower price than bread made from superior flour, but it still appeared as being refined. In 1758, the British government banned the use of alum, but it was still in common use in the mid-1800s as it was difficult to detect by chemical tests.

Now, let's advance in time to the 20th century. Following the industrial and chemical revolutions, the complicated old ways of making white bread were superseded by modern machines with rolling mills that could grind wheat berries to a fine powder and simultaneously remove the wheat germ and fiber that contributed color and coarseness. Time intensive sifting was no longer needed. But there was still the problem of color. Even without the germ, most varieties of wheat flour still had a slight yellow color. How to make cheap and consistent white flour so mills could sell tons of it to commercial bakeries? Aging the flour in a high oxygen atmosphere could whiten it, but that process took months. In the early 1900s, chemical bleaching agents or oxidizers were introduced. Today there are a handful of chemicals that turn flour pristine white and at the same time improve the bread-making process. Some of these are azodicarbonamide (ADA), ascorbic acid (Vitamin C), chlorine, chlorine dioxide, benzoyl peroxide, and calcium peroxide. Unfortunately, some of these chemicals are toxic, but, since they are considered processing aids, they don't have to be listed in the ingredients label. Most of the chemicals are volatile, so, theoretically, they should not wind up in the finished products, but you never know ... accidents do happen! All bleaching agents have been banned in the United Kingdom. Several nations, including the European Union, Canada, and China, have banned the use of benzoyl peroxide and other peroxides as processing aids because of health concerns.

So, in the early part of the 20th century the public, rich and poor alike, got easy access to a cheap and plentiful supply of white bread ... it looks good and has a desirable soft texture (think Wonder Bread). Everybody's happy, right? No! Also, early in the 20th century, biochemists and doctors discovered an alarming link between nutritional deficiencies and deadly diseases. As mentioned above, two diseases, pellagra and beriberi, were diet related. With the discovery of vitamins and their role in human health, the mystery of those diseases was solved. Pellagra was due to a deficiency in vitamin B3, while beriberi was due to a deficiency in Vitamin B1. As scientists learned, in the processing of white flour and white rice, the B vitamins get removed, along with a bunch of other nutrients, like fiber. So, here is truly one of the great ironies and follies of human existence. People, by virtue of their personal preferences, desire refined foods that cause vitamin deficiencies leading to diseases. Now, if humans were really a sensible species and interested in their wellbeing, then you would think that people would want to give up eating white bread, which was bad for health, and return to eating whole wheat bread, which was good for health. People would just go back to eating the darker, unbleached, whole-grain bread with its full complement of vitamins, minerals, fiber, and other goodies. Problem solved! But, no! As mentioned above, the United States government in the 1940s turned to enrichment of white flour using synthetic versions of vitamins and minerals. Note that the later addition of folate (or folic acid) to bread flour is not called enrichment but falls in the category of fortification (adding a micronutrient that wasn't originally present in the food). It's a micronutrient added to food to increase its content to levels greater than naturally found. Another example of fortification would be calcium added to milk.

Further Readings:
Food Crumbles, *A Short History of (White) Bread in the USA*
https://tinyurl.com/2p8hdmu4
The Spruce Eats, "Whole Wheat vs. White Flour"
https://www.thespruceeats.com/whole-wheat-flour-vs-white-flour-2238373
Swindled: *The Dark History of Food Fraud* by Bee Wilson
https://tinyurl.com/243vu6k8

Pizza Hut's Tuscani® Creamy Chicken Alfredo Pasta (990 Calories)

Ingredients:

Rotini Pasta (Water(1), Enriched Semolina (Durum Wheat Semolina(2), Niacin(3), Ferrous Sulfate(4), Thiamine Mononitrate(5), Riboflavin(6), Folic Acid(7))), Alfredo Sauce (Skim Milk(8), Water(9), Soybean(10) or Canola(11) Oils, Parmesan and Romano Cheeses (Part Skim Cow's Milk(12), Enzymes(13), Salt(14), Cellulose(15)), Modified Corn Starch(16), Contains 2% or Less of: Salt(17), Dried Cream (Cream(18), Nonfat Milk(19), Tocopherols(20) and Ascorbyl Palmitate(21) (To Help Protect Flavor), Butter (Cream(22), Natural Flavor(23), Salt(24), Parmesan Cheese Paste (Parmesan Cheese (Milk(25), Cheese Cultures(26), Salt(27), Enzymes(28)), Water(29), Salt(30), Sodium Phosphate(31), Sodium Citrate(32)), Garlic Puree(33), Spice(34), Cheese Flavor (Cheddar Cheese (Cultured Milk(35), Salt(36), Enzymes(37)), Water(38), Salt(39), Enzymes(40), Cultures(41), Phosphoric Acid(42), Xanthan Gum(43), Potassium Sorbate(44)), Enriched Flour (Bleached Wheat Flour(45), Niacin(46), Reduced Iron(47), Thiamine Mononitrate(48), Riboflavin(49), Folic Acid(50), DATEM(51), Mono- & Diglycerides(52), Yeast Extract(53), Xanthan Gum(54), Natural Flavors(55), Lactic Acid(56), Calcium Lactate(57), Chicken (White Meat(58), Water(59), Seasoning (Salt(60), Yeast Extract(61), Spices(62), Dried Cane Syrup(63), Dextrose(64), Carrageenan(65), Dried Chicken Broth(66), Dried Garlic(67), Dried Onion(68), Chicken Fat(69), Dried Parsley(70)), Modified Food Starch(71), Sodium Phosphates(72)), Cheese (Part Skim Mozzarella Cheese: (Pasteurized Milk(73) and Skim Milk(74), Cheese Cultures(75), Salt(76), Enzymes(77)), Sugar Cane Fiber(78) (Added to Prevent Clumping), Modified Food Starch(79), Potassium Chloride(80), Natural Flavors(81), Rosemary Extract(82) (To Protect Flavor).

Ingredient Assignments and the Degree of Industrialization (Processing)

Ascorbyl Palmitate (synthetic) (21) Extremely; Bleached Wheat Flour (45) Highly; Calcium Lactate (synthetic) (57) Extremely; Canola Oil (11) Highly; Carrageenan (65) Moderately; Cellulose (15)Moderately; DATEM (synthetic) (51) Extremely; Dextrose (synthetic) (64) Extremely; Dried Cane Syrup (synthetic) (63) Extremely; Enzymes (13,28,37,40,77) Highly; Ferrous Sulfate (synthetic) (4) Extremely; Folic Acid (synthetic) (7,50) Extremely; Lactic Acid (synthetic) (56) Extremely; Modified Corn Starch (16) Extremely; Modified Food Starch (71,79) Extremely; Mono- & Diglycerides (synthetic) (52) Extremely; Natural Flavors (23,55,81) Moderately; Niacin (synthetic) (46) Extremely; Phosphoric Acid (synthetic) (42) Extremely; Potassium Chloride (synthetic) (80) Extremely; Potassium Sorbate (synthetic) (44) Extremely; Reduced Iron (synthetic) (47) Extremely; Riboflavin (synthetic) (6,49) Extremely; Rosemary Extract (82) Highly; Sodium Citrate (synthetic) (32) Extremely; Sodium Phosphate (synthetic) (31,72) Extremely; Soybean Oil (10) Highly; Sugar Cane Fiber (78) Highly; Thiamine Mononitrate (synthetic) (5,48) Extremely; Tocopherols (synthetic) (20) Extremely; Xanthan Gum (synthetic) (43,54) Extremely.

Reporting the Processed Food Index (PFI) for Creamy Chicken Alfredo Pasta

# OF UNPROCESSED INGREDIENTS:	5
# OF LIGHTLY PROCESSED INGREDIENTS:	14
# OF MODERATELY PROCESSED INGREDIENTS:	3
# OF HIGHLY PROCESSED INGREDIENTS:	6
# OF EXTREMELY PROCESSED INGREDIENTS:	22

TOTAL # OF INGREDIENTS:	82
# OF UNIQUE INGREDIENTS:	50 (61%)
PROCESSED FOOD INDEX:	52 (Highly Industrialized)

Commentary

At first reading, a creamy alfredo pasta dish sounds pretty basic, but the Pizza Hut preparation requires 50 unique ingredients. Of these 50 ingredients, 31 of them are categorized as moderately to extremely processed. The Processed Food Index score is 52, significantly high. Of these 50 ingredients, 19 of them are synthetic (industrially made) and would not likely be found in your local grocery store or kitchen.

Just for fun, I found an internet recipe for Betty Crocker Chicken Alfredo Pasta. It has only **10** ingredients! Included are butter, garlic, chicken broth, ziti pasta, salt, pepper, whipping cream, parmesan cheese, chicken, and sweet peas. Why does it take 50 ingredients to make a Pizza Hut version? Keep in mind that the Pizza Hut dish must have the following properties of a processed food since it is designed to be served in thousands of restaurants across the country to millions of people: (1) The dish must be uniform and consistent, so, no matter which Pizza Hut restaurant serves it, the customer gets the same experience; (2) The preparation must be shelf stable since each restaurant does not make it from scratch but uses a mix obtained from a central distribution center;

and (3) This food item has to be well preserved to prevent safety issues, such as food borne illnesses, and potential run-ins with federal agencies like the United States Department of Agriculture or the Food & Drug Administration. Product appeal, consistency in taste and texture, stability over time and transit, and safety are the hallmarks of a fast-food product. All these necessities require a boatload of ingredients to meet customer expectations and to protect the interests of the manufacturer.

Note the calorie count for this pasta dish. A single serving provides a 990-calorie punch. A drink, side, or dessert would add more calories. The pasta dish alone provides almost half the daily recommended calories for an average American. Eating several fast-food meals like this one in a single day would quickly lead to excessive calorie consumption and weight gain.

The highlighted ingredients in Pizza Hut's Tuscani® Creamy Chicken Alfredo Pasta are:

- Ascorbyl Palmitate

- Cellulose

- Sugar Cane Fiber

- Tocopherols

Ascorbyl Palmitate:

The presence of this synthetic chemical is actually a good sign for the fast-food industry. A little background information is necessary to explain that statement. Fats and oils, either animal or plant based, are common ingredients in fast foods. A frequent problem with fats and oils is that they can go rancid. They can react with oxygen in the air to form derivatives that create off-flavors and off-odors in foods. Therefore, additives called anti-oxidants are included in food formulations to remove oxygen and, thus, limit or eliminate rancidity. In the not-so-distant past, two of the most common anti-oxidants were BHA (butylated hydroxyanisole) and BHT (butylated hydroxytoluene). Over time, those additives developed bad reputations. BHA was linked to cancer, allergies, and liver and kidney problems. BHT was linked to the latter two medical issues.

Along comes ascorbyl palmitate to save the day. It's derived from ascorbic acid, a powerful and natural anti-oxidant, commonly known as Vitamin C. The derivative is oil and fat soluble making it a safe and effective replacement for BHA or BHT, so that makes ascorbyl palmitate a better additive for the food industry. However, there can be a downside (isn't there always one!). Palmitate is sourced from palmitic acid which is derived from palm oil. The destruction of rain forests in Southeast Asia in order to plant palm oil trees is environmentally devastating affecting local ecosystems and diminishing populations of threatened species like orangutans.

You may see ascorbyl palmitate on the labels of skin products as well, where it helps to smooth and improve skin health.

Cellulose:

This additive is derived from plants. Common sources for the food industry would be cotton and wood. It provides the rigidity and sturdiness for plants to stand upright. Think of the stiffness of celery stalks. It's a fibrous material and easily absorbs water. It's a source of food for ruminant animals, like cows, but for humans, who can't digest it, cellulose acts as an insoluble fiber. Cellulose comes in various forms, so it may also be seen on labels as cellulose powder or microcrystalline cellulose (MCC). In food, inexpensive cellulose offers a variety of uses such as adding creaminess, increasing firmness, retards spoilage, serves as a binder in meat, a filler, and can be used as a coating to keep food from sticking together. Cellulose powder probably helps to incorporate air in baked goods. The general uses are as a texturizer, an anti-caking agent, a fat substitute, an emulsifier, an extender, and a bulking agent in food production.

In the pasta dish, this additive is found in the Alfredo sauce where it probably serves as a texturizer.

Sugar Cane Fiber:

This ingredient is pretty odd. It's found in the mozzarella cheese portion of the dish. When sugar cane is refined to produce sugar, after the extraction step, a fibrous residue remains consisting mainly of cellulose and a related substance called hemicellulose. This mixture is cleaned up via a multi-step process to produce purified sugar cane fiber. Given the massive amounts of white sugar used by the processed food industry, this by-product, sugar cane fiber, turns out to be a cheap additive to boost fiber in processed foods.

Fiber is a nutrient critical to healthy digestion and appears on Nutrition Facts Labels under the carbohydrate section. In 2016, the FDA, in a nod to the processed food industry, gave official recognition to 7 substances that could be added to foods to boost fiber content and contribute other properties. The list included cellulose, pectin, psyllium husk among others. In 2018, the list was expanded with 8 additional fiber-like materials. One of those was a generic category called mixed plant cell wall fibers, which included sugar cane fiber.

Why is sugar cane fiber found in cheese? There's a description right there in the ingredient list: it helps to prevent clumping. The cheese manufacturer probably ships mozzarella cheese as grated crumbles, so keeping those crumbles from sticking together helps in the preparation of the pasta dish.

Tocopherols:

The various chemical forms of Vitamin E are in a class called tocopherols. They are natural anti-oxidants and fat-soluble, so they can assist in keeping fats and oils from going rancid. Tocopherols are partners with ascorbyl palmitate as preservatives in foods with fats and oils. [Refer to the description for ascorbyl palmitate above for more details.]

Pizza Hut's Super Supreme Personal Pan Pizza® Slice (210 Cal)

Ingredients:

BEEF: Beef(1), Water(2), Salt(3), Tomato Paste(4), Natural Flavors(5), Dextrose(6), Dehydrated Onion(7), Spice(8).

PEPPERONI: Pork(9), Beef(10), Salt(11), Contains 2% or less of: Spices(12), Dextrose(13), Lactic Acid Starter Culture(14), Natural Spice Extractives(15), Extractives of Paprika(16), Extractive of Rosemary(17), Sodium Nitrite(18).

SEASONED PORK: Pork(19), Water(20), Salt(21), Spices(22), Sugar(23), Corn Syrup Solids(24), Autolyzed Yeast Extract(25), Natural Flavors(26).

CLASSIC MARINARA: Tomato Puree (Water(27), Tomato Paste(28)), Maltodextrin(29), Contains 2% or less of the following: Salt(30), Spices(31), Garlic Powder(32), Tomato Fibers(33), Olive Oil(34), Canola Oil(35), Citric Acid(36), Natural Flavors(37).

CHEESE: Mozzarella Cheese (Pasteurized Milk(38), Cheese Cultures(39), Salt(40), Enzymes(41)), Modified Food Starch(42), Sugar Cane Fiber(43), Potassium Chloride(44), Natural Flavors(45), Ascorbic Acid(46) To Protect Flavor.

PERSONAL PAN CRUST: Enriched Flour (Bleached Wheat Flour(47), Malted Barley Flour(48), Niacin(49), Ferrous Sulfate(50), Thiamine Mononitrate(51), Riboflavin(52), Folic Acid(53)), Water(54), Yeast(55), Contains 2% or less of: Salt(56), Soybean Oil(57), Sugar(58), Enzymes(59), DATEM(60).

FRESH GREEN PEPPERS: Green Bell Peppers(61).

FRESH MUSHROOM: Button Mushrooms (62).

FRESH RED ONIONS: Red Onions(63).

ITALIAN SAUSAGE: Pork(64), Seasoning (Spices(65) (Including Mustard Seed(66)), Salt(67), Paprika(68), Garlic Powder(69), Sugar(70), Natural Flavors(71), Extractives of Paprika(72)), Water(73).

MEDITERRANEAN BLACK OLIVES: Ripe Olive Wedges(74), Water(75), Salt(76), Ferrous Gluconate(77).

SLOW-ROASTED HAM: Ham(78) (Cured with Water(79), Salt(80), Sodium Lactate(81), Sugar(82), Sodium Phosphates(83), Sodium Diacetate(84), Sodium Erythorbate(85), Sodium Nitrite(86).

Ingredient Assignments and the Degree of Industrialization (Processing)

Ascorbic Acid (synthetic) (46) Extremely; **Autolyzed Yeast Extract** (25) Highly; **Bleached Wheat Flour** (47) Highly; **Canola Oil** (35) Highly; **Citric Acid** (synthetic) (36) Extremely; **Corn Syrup Solids** (synthetic) (24) Extremely; **DATEM** (synthetic) (60) Extremely; **Dextrose** (synthetic) (6,13) Extremely; **Enzymes** (41,59) Highly; **Extractives of Paprika** (16,72) Highly; **Extractive of Rosemary** (17) Highly; **Ferrous Gluconate** (synthetic) (77) Extremely; **Ferrous Sulfate** (synthetic) (50) Extremely; **Folic Acid** (synthetic) (53) Extremely; **Malted Barley Flour** (48) Moderately; **Maltodextrin** (29) Highly; **Modified Food Starch** (42) Extremely; **Natural Flavors** (5,26,37,45,71) Moderately; **Natural Spice Extractives** (15) Highly; **Niacin** (synthetic) (49) Extremely; **Olive Oil** (34) Highly; **Potassium Chloride** (synthetic) (44); Extremely; **Riboflavin** (synthetic) (52) Extremely; **Sodium Diacetate** (synthetic) (84) Extremely; **Sodium Erythorbate** (synthetic) (85) Extremely; **Sodium Lactate** (synthetic) (81) Extremely; **Sodium Nitrite** (synthetic) (18,86) Extremely; **Soybean Oil** (57) Highly; **Sodium Phosphates** (synthetic) (83) Extremely; **Sugar** (23,70,82) Moderately; **Sugar Cane Fiber** (43) Highly; **Thiamine Mononitrate** (synthetic) (51) Extremely.

Reporting the Processed Food Index (PFI) for Super Supreme Pizza

# OF UNPROCESSED INGREDIENTS:	10
# OF LIGHTLY PROCESSED INGREDIENTS:	8
# OF MODERATELY PROCESSED INGREDIENTS:	3
# OF HIGHLY PROCESSED INGREDIENTS:	11
# OF EXTREMELY PROCESSED INGREDIENTS:	18

TOTAL # OF INGREDIENTS:	86
# OF UNIQUE INGREDIENTS:	50 (58%)
PROCESSED FOOD INDEX:	48 (Highly Industrialized)

Commentary

Of course, the centerpiece of the Pizza Hut menu is the pizza, so, this entry and the next two, are examples of the many Pizza Hut offerings. Since pizzas can get pretty complex with all the options for toppings, we can expect that there would a significant number of ingredients. And that's what we see in the Super Supreme pizza. There are a total of 86 ingredients with 50 of them being unique. If you wanted to make a similar meal at home, here's what you'd need using store-bought ingredients based on a Betty Crocker recipe (the number of ingredients per item are in parentheses):

- Home-made Crust (5)
- Water (1)
- Vegetable Oil (1)
- Bridgford Pepperoni (12)
- Supermarket Beef (1)
- Pederson's Uncured Sausage (9)

- Esposito's Italian Sausage (9)

- Burger's Smoked City Ham (11)

- Whole Foods 365 Mozzarella Cheese (4)

- Botticelli Marinara Sauce (7)

- Green Peppers (1)

- Mushrooms (1)

- Red Onions (1)

- Whole Foods 365 Black Olives (3)

The total number of ingredients for this home-made pizza is 66 ingredients. That's still a bunch of ingredients, but 20 less than the Pizza Hut version. Most likely, dishes from fast food restaurants will always have more ingredients in them because of the need for uniformity and preservation. Whether you're making dinner at home using highly processed ingredients or getting it as carry out, the product will likely be highly industrialized. The Processed Food Index for the Super Supreme pizza is 48, a pretty high score.

The highlighted ingredients in Pizza Hut's Super Supreme Personal Pan Pizza® Slice are:

- Ferrous Gluconate

- Ferrous Sulfate

- Sodium Diacetate

- Sodium Erythorbate

- Sodium Nitrite

Ferrous Gluconate (Iron (II) Gluconate):

This additive is a chemical compound of iron and gluconic acid (an oxidized form of glucose). It's a black powder. The ferrous part of the name is derived from "ferrum" which is Latin for iron (the chemical symbol for iron, Fe, comes from the Latin name). There are two forms of compounded iron, ferrous and ferric. Only the former type is reactive toward oxygen.

Ferrous gluconate is found in the preparation of the black olives used as a topping for the pizza. Unripe fruits on an olive tree are green, but, given enough time, the fruits will eventually turn black. The green olives are bitter, so, to improve the taste, they need to be cured. Green olives are cured in salt water (brine) over time. However, the process of forming black olives can be speeded up by exposing the green olives to an alkaline solution containing lye, compressed air, and ferrous gluconate, which transforms the skin of the green olive into a shiny, black color. There are subtle differences in taste, appearance, and texture between green and black olives, and each type has its afficionados. Green olives offer more health benefits, but the consumption of black olives supplements the body with iron.

Ferrous Sulfate:

This iron compound shows up in the pizza crust. It's used in the enrichment of flour to provide a source of the mineral iron. Ferrous sulfate is just one of several forms of iron that the grain milling industry uses to enrich refined wheat flour. Other iron sources that may appear in an ingredients list for flour include ferrous gluconate and reduced iron (another name for ferrous). Sometimes the type of iron is not specified and you'll just see "iron" listed, which is deceptive. Since iron is a metal, it's not readily digestible, so a compounded form of it must be used for enrichment.

Sodium Diacetate (Dry Acetic Acid):

This synthetic additive is an ingredient in the slow-roasted ham, which is used as a meat topping for the pizza. Sodium diacetate has several applications in the food industry, but, primarily, it functions as fungicide (mold inhibitor) and bactericide (kills bacteria). It's a white, crystalline solid that readily absorbs water (hygroscopic) and has the smell of acetic acid (main ingredient in vinegar). The compound is made from the combination of acetic acid and sodium acetate.

In the slow-roasted ham, low levels of sodium diacetate (under 0.2%) decreases pH (higher acidity) on the surface of the ham and inhibits the growth of mold and bacteria, which could lead to foodborne pathogens, like listeria.

Other applications for sodium diacetate include flavoring foods by imparting a vinegary taste and to serve as a buffer to adjust acidity.

Sodium Erythorbate:

This synthetic chemical is another preservative added to the slow-roasted ham. It's a white, odorless powder. It works in tandem with sodium nitrite (see next entry) to cure meat and produce the characteristic and appealing red/pink color of processed meats. It serves as an antioxidant (maintains color and flavor), a preservative (extends shelf life by inhibiting microbial growth), and accelerates the curing process of meat. As a side-benefit, it helps to minimize the detrimental health effects of sodium nitrite. This additive is increasingly being used as a substitute for sulfite preservatives since the latter can cause allergic reactions in some people.

Sodium Nitrite:

The claim to fame of this chemical is that it's one of the most controversial additives to be introduced by food manufacturers. In the Pizza Hut pizza, it's listed as an ingredient in the pepperoni and slow-roasted ham. In the early 1900s, sodium nitrite was discovered to be an ideal curing agent for meats, where it could convert the unappealing brown color of meat into a fresh-looking pink color. Other techniques had been used in the past, but they were not as effective. For example, in the 1800s, potassium nitrate (saltpeter), was the primary curing agent, but its action took weeks as opposed to hours in the case of sodium nitrite. Germany, after WWI, started selling sodium nitrite to other countries since they had an overstock of the chemical, which had been used for munition manufacturing during the war. It was sold under different names, such as Prague Salt or Curing Salt. In order not to cause confusion with table salt, the material was dyed pink. As a side benefit, food scientists found that sodium nitrite also kills the very toxic botulism bacteria. With this inexpensive and effective curing agent, the processed meat industry in the US took off in the 1920s.

The World Health Organization (WHO) in 2015, based on a meta-analysis of over 800 studies, linked processed meats to a significant risk of human cancer, particularly colorectal cancer. A positive association was made with stomach cancer. Intake of processed meats also contributed to increased risks of coronary heart disease, stroke, and Type 2 diabetes. Although the exact causes of cancer are not known, sodium nitrite is considered a culprit. The FDA considers the additive safe at levels of 200 ppm in processed meats (Code of Federal Regulations 172.175). That value equates to 0.09g of sodium nitrite per pound of meat. Sodium nitrite is toxic in humans at a level equal to or exceeding 32mg per pound of body weight. For a 150-lb person, that value equates to 4.8g. If the 4.8g is divided by 0.09g of sodium nitrate per pound of meat, the result is 53 lbs of meat. This calculation suggests that a 150-lb person would have to eat 53 lbs of processed meats in a single day to reach a toxic level of sodium nitrite. The FDA considers that level of meat consumption to be improbable, and hence they deem the 200-ppm maximum amount of sodium nitrite to be safe for human consumption. However, that calculation applies only to sodium nitrite and not to processed meats in general which carry additional risks associated with their ingredients or cooking methods.

Further Readings:
Earthworm Express: *Nitrate and Nitrite – Their History and Functionality*

https://tinyurl.com/9bsncjjt

Harvard School of Public Health: *"WHO Report Says Eating Processed Meat Is Carcinogenic"*

https://tinyurl.com/2ns358jr

P. H.'s Meat Lover's® Original Stuffed Crust® Slice (430 Calories)

Ingredients:

APPLEWOOD SMOKED BACON: Bacon(1) (Cured with Water(2), Salt(3), Sugar(4), Sodium Phosphates(5), Sodium Erythorbate(6), Sodium Nitrite(7)), May Contain: Smoke Flavoring(8), Brown Sugar(9).

SLOW-ROASTED HAM: Ham(10) (Cured with Water(11), Salt(12), Sodium Lactate(13), Sugar(14), Sodium Phosphates(15), Sodium Diacetate(16), Sodium Erythorbate(17), Sodium Nitrite(18).

BEEF: Beef(19), Water(20), Salt(21), Tomato Paste(22), Natural Flavors(23), Dextrose(24), Dehydrated Onion(25), Spice(26).

ITALIAN SAUSAGE Pork(27), Seasoning (Spices (Including Mustard Seed(28), Salt(29), Paprika(30), Garlic Powder(31), Sugar(32), Natural Flavors(33), Extractives of Paprika(34)), Water(35).

PEPPERONI: Pork(36), Beef(37), Salt(38), Contains 2% or less of: Spices(39), Dextrose(40), Lactic Acid Starter Culture(41), Natural Spice Extractives(42), Extractives of Paprika(43), Extractive of Rosemary(44), Sodium Nitrite(45).

SEASONED PORK: Pork(46), Water(47), Salt(48), Spices(49), Sugar(50), Corn Syrup Solids(51), Autolyzed Yeast Extract(52), Natural Flavors(53).

CLASSIC MARINARA: Tomato Puree (Water(54), Tomato Paste(55)), Maltodextrin(56), Contains 2% or less of the following: Salt(57), Spices(58), Garlic Powder(59), Tomato Fibers(60), Olive Oil(61), Canola Oil(62), Citric Acid(63), Natural Flavors(64).

CHEESE: Mozzarella Cheese (Pasteurized Milk(65), Cheese Cultures(66), Salt(67), Enzymes(68)), Modified Food Starch(69), Sugar Cane Fiber(70), Potassium Chloride(71), Natural Flavors(72), Ascorbic Acid(73) To Protect Flavor.

ORIGINAL STUFFED CRUST: Enriched Flour (Bleached Wheat Flour(74), Malted Barley Flour(75), Niacin(76), Ferrous Sulfate(77), Thiamine Mononitrate(78), Riboflavin(79), Folic Acid(80)), Water(81), Yeast(82), Soybean Oil (83) Contains 2% or less of: Salt(84), Vital Wheat Gluten(85), DATEM(86), Sugar(87), Enzymes(88), Ascorbic Acid(89), Sucralose(90), Soybean Oil(91), TBHQ(92), Mozzarella Cheese (Pasteurized Milk(93), Cheese Cultures(94), Salt(95), Potassium Chloride(96), Enzymes(97)), Modified Food Starch(98), Nonfat Milk(99), Flavors(100).

Buttery Blend Crust Flavor (Medium or Large Pizza): Soybean Oil(101), Water(102), Salt(103), Vegetable Mono & Diglycerides(104), Contains Less Than 2% Of: Sodium Benzoate(105) and Potassium Sorbate(106) (As Preservatives), Soy

Lecithin(107), Citric Acid(108), Natural Flavor(109), Calcium Disodium EDTA(110) Added To Protect Flavor, Beta Carotene(111) (Color), Vitamin A Palmitate(112) Added.

Ingredient Assignments and the Degree of Industrialization (Processing)

Ascorbic Acid (synthetic) (73,89) Extremely; **Autolyzed Yeast Extract** (52) Highly; **Beta Carotene** (111) Highly; **Bleached Wheat Flour** (74) Highly; **Brown Sugar** (9) Moderately; **Calcium Disodium EDTA** (synthetic) (110) Extremely; **Canola Oil** (62) Highly; **Citric Acid** (synthetic) (63,108) Extremely; **Corn Syrup Solids** (synthetic) (51) Extremely; **DATEM** (synthetic) (86) Extremely; **Dextrose** (synthetic) (24) Extremely; **Enzymes** (68,88,97) Highly;

Extractives of Paprika (34,43) Highly; **Extractive of Rosemary** (44) Highly; **Ferrous Sulfate** (synthetic) (77) Extremely; **Flavors** (synthetic) (100) Extremely; **Folic Acid** (synthetic) (80) Extremely; **Malted Barley Flour** (75) Moderately; **Maltodextrin** (56) Highly; **Modified Food Starch** (69,98)Extremely; **Natural Flavors** (23,53,64,72,109) Moderately; **Niacin** (synthetic) (76) Extremely; **Olive Oil** (61) Highly; **Potassium Chloride** (synthetic) (71,96) Extremely; **Potassium Sorbate** (synthetic) (106) Extremely; **Riboflavin** (synthetic) (79) Extremely; **Smoke Flavoring** (8) Highly; **Sodium Benzoate** (synthetic) (105) Extremely; **Sodium Diacetate** (synthetic) (16) Extremely; **Sodium Erythorbate** (synthetic) (6,17) Extremely; **Sodium Lactate** (synthetic) (13) Extremely; **Sodium Nitrite** (synthetic) (7,18,45) Extremely; **Sodium Phosphates** (synthetic) (5,15) Extremely; **Soy Lecithin** (107) Highly; **Soybean Oil** (83,91,101) Highly; **Sucralose** (synthetic) (90) Extremely; **Sugar** (4,14,87) Moderately; **Sugar Cane Fiber** (70) Highly; **TBHQ** (synthetic) (92) Extremely; **Thiamine Mononitrate** (synthetic) (78) Extremely; **Vegetable Mono &**

Diglycerides (synthetic) (104) Extremely; **Vital Wheat Gluten** (85) Moderately; **Vitamin A Palmitate** (synthetic) (112) Extremely.

Reporting the Processed Food Index (PFI) for Meat Lover's® Pizza

# OF UNPROCESSED INGREDIENTS:	7
# OF LIGHTLY PROCESSED INGREDIENTS:	9
# OF MODERATELY PROCESSED INGREDIENTS:	5
# OF HIGHLY PROCESSED INGREDIENTS:	13
# OF EXTREMELY PROCESSED INGREDIENTS:	25

TOTAL # OF INGREDIENTS:	112
# OF UNIQUE INGREDIENTS:	59 (53%)
PROCESSED FOOD INDEX:	55 (Highly Industrialized)

Commentary

This pizza option has an abundance of ingredients at 112. Over half of the ingredients (53%) are unique. If you wanted to make a similar meal at home, here's what you'd need using store-bought ingredients based on a Betty Crocker recipe (the number of ingredients per item are in parentheses) for a stuffed-crust pizza with meat toppings:

- 1/2-pound bulk Italian sausage, Johnsonville (9)
- 1/2-pound lean ground beef (1)
- 3 1/3 cups Original Bisquick™ mix (8)
- 3/4 cup cold water (1)
- 3 cups shredded mozzarella cheese (12 ounces), Kraft (6)
- 1 jar (14 to 15 ounces) pizza sauce, Kraft (16)
- 1 cup sliced fresh mushrooms (3 ounces) (1)
- 1/4 cup chopped green bell pepper (1)

The total number of ingredients for this home-made version is 43. That's still a bunch of ingredients, but 69 less than the Pizza Hut version. Invariably, dishes from fast food restaurants will always have more ingredients in them because of the need for uniformity and preservation. Whether you're making dinner at home using highly processed ingredients or getting it as carry out, the product will still be highly industrialized. The Processed Food Index for the Meat Lover's® Pizza is 55, a pretty high score.

The highlighted ingredients in Pizza Hut's Meat Lover's® Original Stuffed Crust® Slice are:

- Beta Carotene

- Corn Syrup Solids

- Sucralose

Beta Carotene:

This additive is found in the buttery blend crust flavor. As noted in the ingredient list, beta carotene is added for color. Why is that? Since the name of the crust includes the word "buttery" but there is no butter in the composition of the crust, a food coloring is added to give the appearance of a buttery crust. Beta carotene in its pure state is a red-orange color.

The use of beta carotene for color is a positive sign for the processed food industry, since, in the past, a company would have used artificial dyes with potential health risks (see section on McDonald's McCafé® Mocha Frappé). Since beta carotene is present in natural food sources, such as orange vegetables (carrots) and orange fruits (cantaloupes), it carries no safety concerns. It belongs to a class of organic compounds aptly named carotenoids. Actually, this pigment offers health benefits, since it is a precursor for the biosynthesis of Vitamin A. In terms of the degree of processing, beta carotene is labeled as highly processed. It is not synthetic, but its isolation and purification require the extraction of a rich-source, such as algae, using toxic solvents like hexane.

Additionally, beta carotene, as an anti-oxidant, can protect against oxidative stress in the body, which could lead to serious health problems like cancer, heart disease, and dementia.

Corn Syrup Solids (Corn Syrup Powder):

We know that corn syrup is a liquid sweetener. If water is removed (dehydration), corn syrup can be crystallized into a solid sweetener. In the formulation of some processed foods, it is more advantageous to add a solid sweetener rather than a liquid one. In the pizza toppings, corn syrup solids are found in the seasoned pork. Why is sugar added to pork? According to meat science, a sweetener is added to cured meats to (1) counteract the intense saltiness, (2) aid in the bacterial fermentation process that occurs during curing, and (3) assist in maintaining color.

Sucralose:

This ingredient is found in the stuffed crust and adds sweetness to it. As a high-intensity sweetener (HIS), you need very little of it in a formula, since it is 320+ times sweeter than table sugar. It is 3 times as sweet as the HISs aspartame and acesulfame potassium and twice as sweet as saccharin. This synthetic chemical is fairly stable, but may break down at temperatures exceeding 248°F (that's lower than pizza baking temperatures!).

The history of sucralose is pretty fascinating. It was discovered by accident in 1976 by two chemists (Leslie Hough and Shashikant Phadnis) working for the Tate & Lyle Company in England. They were researching unique derivatives of sucrose. One derivative was made using a chlorinating agent (a reagent that adds chlorine atoms to another chemical). When one chemist asked the other to "test" the compound, the other researcher thought he said "taste." So, taste he did and found out how extraordinarily sweet the derivative was. Tate & Lyle patented the substance and by 1998, sucralose received approval from the FDA as an additive in foods and beverages. The product was sold and distributed by McNeil Nutritionals, a subsidiary of Johnson & Johnson, under the Splenda brand. The chemical name for sucralose is 1,6-dichloro-1,6-dideoxyfructose–4-chloro-4-deoxygalactose disaccharide. Try saying that phrase five times quickly! One of the chlorinating agents used to make sucralose is phosgene, a chemical weapon first used by the French in World War I.

As an analytical chemist for Tate & Lyle, I became very familiar with sucralose, and tested it in a variety of ways. I would describe sucralose as one of the most industrialized food additives in the world. I don't usually get into the details of manufacturing, but I thought it would be educational to present one of the several ways that sucralose is synthesized in a chemical plant. In that way, you'll have an idea how complicated chemical syntheses can be for some food additives. There's a world of chemistry involved in the processed food industry. Here's an example synthesis:

1. Sucrose (table sugar) is dissolved in dimethyl formamide (hazardous organic solvent) at room temperature. Slowly trimethyl orthoacetate and para-toluene sulfonic acid is added.

2. The mixture is stirred for 2½ hours.

3. Then distilled water is added and the stirring continued for another 40 minutes.

4. Next, tertiary butylamine is added and the mixture stirred for 1½ hours.

5. Some solvent is stripped off to concentrate the mixture.

6. Then ethyl acetate and methanol are added to precipitate sucrose-6-acetate which is mechanically separated from the mixture and purified.

7. The sucrose-6-acetate is dissolved in dimethyl formamide under ice-cold conditions, then thionyl chloride dissolved in 1,1,2-trichloroethane is added dropwise.

8. The mixture is maintained at low temperatures for 30 minutes.

9. The mixture is slowly heated up to 239°F over 2 hours, then boil the mixture for 1½ hours.

10. The mixture is cooled to below 59°F, then dilute ammonia is added dropwise, while the temperature is kept below 86°F until the mixture is neutralized.

11. The organic phase (upper layer) is separated from the aqueous phase (lower phase).

12. The solvent is evaporated under reduced pressure until a thick product (chlorinated sucrose-6-acetate) is obtained.

13. The chlorinated sucrose-6-acetate is dissolved in methyl alcohol containing sodium methoxide.

14. At room temperature and under reduced pressure, the mixture is stirred for 2 hours.

15. The solid product is filtered, purified with methanol, then decolorized using activated carbon, and finally dried to yield a white powder.

Note that there are a variety of hazardous chemicals used throughout this process. However, sucralose can be purified to produce a very clean, non-contaminated product.

Since sucralose is a sweet, white powder and chemically derived from sugar, the Splenda packaging originally had the phrase "made from sugar so it tastes like sugar." A trade group of the sugar industry, the Sugar Association, filed a complaint with the Federal Trade Commission (FTC) claiming that the marketing of sucralose by McNeil Nutritionals was confusing consumers. The Sugar Association prevailed and the catchy marketing phrase disappeared.

Since the FDA approved sucralose as a sweetener in 1998, it was considered safe to consume. However, there has been controversy about how safe it is. When ingested in the human body, most of the sucralose is not absorbed and winds up in the feces. The fraction (20 to 30%) that is absorbed and metabolized winds up getting processed by the kidneys and eliminated in the urine. Sucralose has not been found to be carcinogenic. However, at high cooking temperatures (>248°F), as in baking, roasting, or deep frying, the sweetener can break down to form potentially carcinogenic substances. Also, adding sucralose to hot foods or beverages can lead to a similar breakdown.

Further Readings:
Modern Applied Science, *Synthesis of Strong Sweetener Sucralose*

Open PDF doc at this website

https://tinyurl.com/2tsj5ct5

NBC News, *Sugar Industry Files Complaint Over Splenda*

https://www.nbcnews.com/id/wbna15533454

Pizza Hut's The Great Beyond Original Pan Pizza® Slice (390 Cal)

Ingredients:

BEYOND ITALIAN SAUSAGE: Water(1), Pea Protein(2), Refined Coconut Oil(3), Expeller-Pressed Canola Oil(4), Natural Flavors(5), Dried Yeast(6), Spices(7), Rice Protein(8), Methylcellulose(9), Chicory Root Fiber(10), Salt(11), Paprika(12), Yeast Extract(13), Niacin(14), Pyridoxine Hydrochloride(15), Thiamine Hydrochloride(16), Riboflavin(17), Folic Acid(18), Cyanocobalamin(19), Sugar(20), Garlic Powder(21), Vinegar(22), Lemon Juice Concentrate(23), Onion Powder(24), Sunflower Lecithin(25), Extracts of Rosemary(26).

CLASSIC MARINARA: Tomato Puree (Water(27), Tomato Paste(28)), Maltodextrin(29), Contains 2% or less of the following: Salt(30), Spices(31), Garlic Powder(32), Tomato Fibers(33), Olive Oil(34), Canola Oil(35), Citric Acid(36), Natural Flavors(37).

CHEESE:

Mozzarella Cheese (Pasteurized Milk(38), Cheese Cultures(39), Salt(40), Enzymes(41)), Modified Food Starch(42), Sugar Cane Fiber(43), Potassium Chloride(44), Natural Flavors(45), Ascorbic Acid(46) To Protect Flavor.

PAN CRUST (Original): Enriched Flour (Bleached Wheat Flour(47), Malted Barley Flour(48), Niacin(49), Ferrous Sulfate(50), Thiamine Mononitrate(51), Riboflavin(52), Folic Acid(53)), Water(54), Yeast(55), Contains 2% or less of: Salt(56), Soybean Oil(57), Vital Wheat Gluten(58), Sugar(59), Enzymes(60), Ascorbic Acid(61), Sodium Stearoyl Lactylate(62). Pan Oil: Soybean Oil(63) with TBHQ(64) added to protect freshness.

DICED ROMA TOMATOES: Fresh Roma Tomatoes(65).

FRESH RED ONIONS: Red Onions(66).

SLICED BANANA PEPPERS: Fresh Banana Peppers(67), Vinegar(68), Water(69), Salt(70), Contains Less Than 2% of: Lactic Acid(71), Malic Acid(72), Calcium Chloride(73), Sodium Benzoate(74), Polysorbate 80(75), Sodium Metabisulfite(76), Turmeric Extract(77).

Ingredient Assignments and the Degree of Industrialization (Processing)

Ascorbic Acid (synthetic) (46,61) Extremely; **Bleached Wheat Flour** (47) Highly; **Calcium Chloride** (synthetic) (73) Extremely; **Canola Oil** (35) Highly; **Citric Acid** (synthetic) (36) Extremely; **Cyanocobalamin** (synthetic) (19) Extremely; **Enzymes** (41,60) Highly; **Expeller-Pressed Canola Oil** (4) Moderately; **Extracts of Rosemary** (26) Highly; **Ferrous Sulfate** (synthetic) (50) Extremely; **Folic Acid** (synthetic) (18,53) Extremely; **Lactic Acid** (synthetic) (71) Extremely; **Malic Acid** (synthetic) (72)Extremely; **Malted Barley Flour** (48) Moderately; **Maltodextrin** (29) Highly; **Methyl Cellulose** (synthetic) (9) Extremely; **Modified Food Starch** (42) Extremely; **Natural Flavors** (5,37,45) Moderately; **Niacin** (synthetic) (14,49) Extremely; **Olive Oil** (34) Highly; **Pea Protein** (2) Moderately; **Polysorbate 80** (synthetic) (75) Extremely; **Potassium Chloride** (synthetic) (44) Extremely; **Pyridoxine Hydrochloride** (synthetic) (15) Extremely; **Refined Coconut Oil** (3) Moderately; **Riboflavin** (synthetic) (17,52) Extremely; **Rice Protein** (8) Highly; **Sodium Benzoate** (synthetic) (74); Extremely; **Sodium Metabisulfite** (synthetic) (76) Extremely; **Sodium Stearoyl Lactylate** (synthetic) (62) Extremely; **Soybean Oil** (57,63) Highly; **Sugar** (20,59) Moderately; **Sugar Cane Fiber** (43) Highly; **Sunflower Lecithin** (25) Moderately; **TBHQ** (synthetic) (64) Extremely; **Thiamine Hydrochloride** (synthetic) (16) Extremely; **Thiamine Mononitrate** (synthetic) (51) Extremely; **Turmeric Extract** (77) Moderately; **Vinegar** (22,68) Moderately; **Vital Wheat Gluten** (58) Moderately.

Reporting the Processed Food Index (PFI) for the Great Beyond Pizza

# OF UNPROCESSED INGREDIENTS:	6
# OF LIGHTLY PROCESSED INGREDIENTS:	10
# OF MODERATELY PROCESSED INGREDIENTS:	10
# OF HIGHLY PROCESSED INGREDIENTS:	9
# OF EXTREMELY PROCESSED INGREDIENTS:	21

TOTAL # OF INGREDIENTS:	77
# OF UNIQUE INGREDIENTS:	56 (73%)
PROCESSED FOOD INDEX:	49(Highly Industrialized)

Commentary

The last several years has brought dramatic changes in the US food industry with the introduction and public acceptance of plant-based meats. What used to be a niche industry, primarily marketed to vegetarians and vegans, has become more mainstreamed. Today, in 2021, many fast-food companies have introduced meat substitutes into their restaurant menus, frequently offering the same sandwiches but with the animal-based patties switched out for a plant-based analog. Given the increasing concerns over climate change, environmental degradation, ethical treatment of livestock, and health concerns, this trend is likely to continue and expand. Commercially prepared cultured meats, derived from the controlled growth of isolated animal cells, are just on the horizon, and, no doubt, will be appearing in fast food restaurants in just a few years. All the innovations in alternative meats bring with them new and unique ingredients. Undoubtedly, the FDA will be expanding its GRAS list to accommodate the new trend. Pizza Hut, with The Great Beyond pizza, has thrown its hat in the ring. The sausage topping is manufactured by the Beyond Meat Company, which uses pea protein to mimic pork (see below). This pizza has a smaller caloric content with lower cholesterol and saturated fat levels.

The highlighted ingredients in Pizza Hut's The Great Beyond Original Pan Pizza® Slice are:

- Methylcellulose
- Pea Protein
- Sodium Metabisulfite
- Sunflower Lecithin

Methylcellulose:

This synthetic chemical, a white powder, is a derivative of cellulose, which was described earlier as a binder or thickener in processed foods. Methyl cellulose is prepared from the cellulose in wood or cotton by treatment with chloromethane under alkaline conditions. This inexpensive material ($23/lb at Amazon) has a multitude of uses in the industrial world. In food, it acts as a binder, thickener, dispersant, and emulsifying agent. It adds creaminess, enhances firmness, retards spoilage, and can be used as a coating to prevent food from sticking together. In the pizza, methylcellulose is an ingredient in the Beyond Meat Italian sausage where, most likely, it acts as a binder to hold the sausage bits together.

Methylcellulose is a wonderful example of a chemical that not only has many uses in the food industry, but also intersects with many other industries. As a bulk-forming and stool-forming agent, it's used in the treatment of constipation. As a water-soluble lubricant, it's used to make artificial tears. It's used to make non-gelatin drug and supplement capsules for vegetarians and vegans. It's a major ingredient in wallpaper pastes. In the movie "**Ghostbusters**," methylcellulose was used to make the supernatural slime.

Pea Protein (Pea Protein Isolate):

The Beyond Meat plant-based Italian sausage gets its protein from peas. Pea protein has advantages over soy protein since it is non-allergenic, non-gmo, and requires less processing. It has a whopping 75% protein, although the amino acids in the protein are not as well balanced as in animal-based protein. Pea protein is easily digestible by most people, and it's a good source of iron.

How is pea protein powder made? It's a bit of a trade secret, but here's a generic processing procedure: (1) Yellow peas are cleaned to remove impurities; (2) The peas are split to separate the hulls (outer shells); (3) The protein fraction in the de-hulled peas is extracted in an alkaline solution; (4) The protein is precipitated from the mixture using an isoelectric method (pH is changed until the protein becomes insoluble); (5) The protein is separated by centrifugation, then washed, neutralized, and de-watered by a spray-drying process. The final result is a beige-colored powder.

Further Readings:
Wikipedia, *"Pea Protein"*

https://en.wikipedia.org/wiki/Pea_protein

Sodium Metabisulfite:

This synthetic, white, crystalline substance has a multitude of uses in many industries, but, in the food industry, it's known as a sulfiting agent. The FDA recognizes 6 sulfiting agents as GRAS: sulfur dioxide, sodium sulfite, sodium and potassium bisulfite, and sodium and potassium metabisulfite. They are all related in terms of their functions.

Sulfiting agents are used in the wine-making industry to inhibit bacterial growth, and they also inhibit the fermentation of sugar. When in excess of 10ppm, the agent must be listed on the label since some people are allergic to it. They serve as preservatives for fruit and vegetable juices (If you have lemon juice in your kitchen, look at its label). Sodium metabisulfite is likely one of the ingredients.) They are often found in dried fruits as a preservative. Also, they are added to maraschino cherries. They are not allowed to be used on <u>fresh</u> fruits and vegetables.

In the pizza dish, sodium metabisulfite is found in the banana peppers, where it serves as a preservative.

Some consumers are sulfite sensitive and need to be careful with eating foods containing sulfiting agents. Reactions to sulfites include acute asthma attacks, anaphylactic shock, diarrhea, nausea, and headaches. Sulfites are not allowed in products with thiamine (vitamin B1) since the vitamin will be destroyed in their presence.

Sunflower Lecithin:

This yellow-brownish fatty material is another ingredient in the Beyond Italian sausage. Lecithin is incorporated in the membrane of every cell of the body. It belongs to a larger group of molecules called phospholipids, which are essential to the brain, blood, nerves, and other body parts. Lecithin is converted to choline, the precursor to the neurotransmitter acetylcholine. Lecithin is also found in other animal tissues (first discovered in egg yolks) as well as plant tissues such as soy and sunflower. Soy used to be the main source of lecithin, but, due to concerns about GMOs (genetically modified organisms), the food industry has been switching to the sunflower variety. In its pure state, lecithin is nasty to work with since, if you get it on your hands, it's difficult to wash off, and yet, if you've ever used a spray lubricant like Pam to keep foods from sticking to a stovetop pan, oddly lecithin is the main ingredient in the can.

Sunflower lecithin is made by dehydrating sunflowers and separating them into three parts: oil, gum, and solids. The lecithin comes from the gum. It is processed through a cold press system like the one used to make olive oil. This ingredient serves as an emulsifying agent, which keeps oil from separating out of mixtures that contain both. That's probably its purpose in the sausage where coconut oil, canola oil, and water are present together.

Before leaving the Beyond Italian Sausage, it's interesting to note that it's enriched with vitamins, similar to the ones in enriched flour. There is no requirement that the Beyond Company nutritionally enrich its product. Actually, there's a vitamin in the mix that is rarely seen as an ingredient in processed foods: cyanocobalamin (Vitamin B12). That's a nod to vegetarians and vegans who, by not consuming meat products, may be prone to a Vitamin B12 deficiency in their diet. There aren't any plant foods that are a good source of Vitamin B12.

Pizza Hut's Cinnabon® Stick (80 Calories)

Ingredients:

CINNABON STICK: Enriched Flour Bleached (Wheat Flour(1), Malted Barley Flour(2), Niacin(3), Ferrous Sulfate(4), Thiamine Mononitrate(5), Riboflavin(6), Folic Acid(7)), Water(8), Yeast(9), Salt(10), Soybean Oil(11), High Fructose Corn Syrup(12), DATEM(13), Vital Wheat Gluten(14), Enzymes(15), Ascorbic Acid(16), Sodium Stearoyl Lactylate(17). Pan Oil: Soybean Oil(18), TBHQ(19) Added To Protect Freshness. Buttery Blend: Soybean Oil(20), Water(21), Salt(22), Vegetable Mono & Diglycerides(23), Contains Less Than 2% Of: Sodium Benzoate(24) And Potassium Sorbate(25) (As Preservatives), Soy Lecithin(26), Natural Flavor(27), Citric Acid(28), Calcium Disodium EDTA(29) Added To Protect Flavor, Beta Carotene(30) (Color), Vitamin A Palmitate(31) Added.

TOPPING: Cinnamon(32) And Sugar(33).

Ingredient Assignments and the Degree of Industrialization (Processing)

Ascorbic Acid (synthetic) (16) Extremely; **Beta Carotene** (synthetic) (30) Extremely; **Calcium Disodium EDTA** (synthetic) (29) Extremely; **Citric Acid** (synthetic) (28)Extremely;

DATEM (synthetic) (13) Extremely; **Enzymes** (15) Highly; **Ferrous Sulfate** (synthetic) (4) Extremely; **Folic Acid** (synthetic) (7) Extremely; **High Fructose Corn Syrup** (synthetic) (12) Extremely; **Malted Barley Flour** (2) Moderately; **Natural Flavor** (27) Moderately; **Niacin** (synthetic) (3)Extremely; **Potassium Sorbate** (synthetic) (25) Extremely; **Riboflavin** (synthetic) (6) Extremely; **Sodium Benzoate** (synthetic) (24) Extremely; **Sodium Stearoyl Lactylate** (synthetic) (17) Extremely; **Soy Lecithin** (26) Highly; **Soybean Oil** (11,18,20) Highly; **Sugar** (33) Moderately; **TBHQ** (synthetic) (19) Extremely; **Thiamine Mononitrate** (synthetic) (5) Extremely; **Vegetable Mono & Diglycerides** (synthetic) (23) Extremely; **Vitamin A Palmitate** (synthetic) (31) Extremely; **Vital Wheat Gluten** (14) Moderately; **Wheat Flour** (1) Moderately.

Reporting the Processed Food Index (PFI) for the Cinnabon® Stick

# OF UNPROCESSED INGREDIENTS:	2
# OF LIGHTLY PROCESSED INGREDIENTS:	1
# OF MODERATELY PROCESSED INGREDIENTS:	5
# OF HIGHLY PROCESSED INGREDIENTS:	3
# OF EXTREMELY PROCESSED INGREDIENTS:	17

TOTAL # OF INGREDIENTS:	33
# OF UNIQUE INGREDIENTS:	28 (85%)
PROCESSED FOOD INDEX:	69 (Highly Industrialized)

Commentary

This item, one of just a handful of desserts at Pizza Hut, is now called "Cinnamon Sticks." When you look at the calorie count of 80 (not including the icing which comes in a separate container), you may think that's pretty low, but they are sold in packs of 10. So, if you eat the whole pack, given that they are designed to look great, smell great, and taste great, 800 calories wind up getting consumed. Compared to menu items previously discussed, the number of ingredients at 33 is pretty small. However, the Processed Food Index is 69, one of the highest of all the items examined in this book. Another way to look at the degree of industrialization is to divide the number of extremely processed ingredients (17) by the total number of unique ingredients (28). Expressed as a percent, the result is 61%. In other words, 61% of the ingredients are man-made and are not considered food in the conventional sense. As expected, the main components of this ultra-processed dessert are flour, sweeteners, fats, and preservatives, with an emphasis on the latter as we shall see shortly.

The highlighted ingredients in Pizza Hut's Cinnabon® Stick are:

- Calcium Disodium EDTA

- Sodium Benzoate and Potassium Sorbate

- TBHQ

Calcium Disodium EDTA:

First of all, notice that this additive name contains an acronym, EDTA. Most consumers would have no idea what that acronym represents, but the FDA allows food manufacturers to use acronyms, probably because of space limitations on labels. Another acronym, TBHQ, will be discussed below. The EDTA stands for ethylene diamine tetraacetic acid. That's a mouthful! Without getting knee-deep into chemistry, the additive is a "salt," a chemical formed from the neutralization reaction of an acid (EDTA) and a base (mixture of calcium and sodium hydroxides). The chemical is a white, crystalline powder with a slightly salty taste.

Calcium Disodium EDTA is a very common food additive found in hundreds of food products. As a preservative, it prevents spoilage, preserves both flavor and color, and increases shelf life. It is also called a chelating agent. Chelating comes from the word "chela," which is Latin for pincer-like claw (as in lobster or crab). Chelating agents have a chemical structure which allows them to grab onto metal ions like zinc, iron, copper, and magnesium. By sequestering (separating) dissolved metals in a food product, the additive can prevent or diminish off-flavors and off-colors from forming. An interesting side note is that Calcium Disodium EDTA is also used in the treatment of lead poisoning (chelation therapy) by removing lead from the bloodstream via the kidneys and excreting it in the urine.

Sodium Benzoate and Potassium Sorbate

Although these two additives are not chemically alike, they are grouped together here because they are both preservatives and often appear together in processed foods.

Sodium Benzoate is a natural product found in fruits like cranberry and bilberry. The additive is a synthetic, white, crystalline powder. As a preservative, it functions as a bactericide and a fungicide. It prevents food spoilage, and it is particularly effective in acidic foods (salad dressings and jams) and beverages (sodas and fruit juices). It's also used as a pickling agent.

Sodium Benzoate is generally considered safe, but in foods that contain vitamin C (ascorbic acid), a chemical reaction can occur leading to the formation of benzene, a carcinogenic substance. In 2005, the FDA found that 10 out of 200 beverages tested contained 5 parts per billion (ppb) of benzene, which is the upper limit for safe drinking water set by the Environmental Protection Agency (EPA). Diet sodas are more prone to form benzene due to the absence of sugar, which mitigates the formation of benzene. Note that the Cinnabon® Stick also contains ascorbic acid, so there's a possibility that benzene could form during digestion in the acidic medium of the stomach.

Potassium Sorbate is probably the most common preservative found in processed foods. You'll find it in cheeses, dips, yogurt, sour cream, baked goods, pies, fermented vegetables, dressings, and beverages, as well as many other foods. The effectiveness of this preservative increases as the acidity of the food increases. It's a white, crystalline, tasteless powder which can inhibit mold, fungi, and yeast. It's derived from sorbic acid, a natural substance, but the industrial version is made using a synthetic method. Potassium sorbate is also used in the wine industry where it inhibits microbial growth after the fermentation process is complete. This additive is very safe and has the lowest allergenic potential of all the food preservatives.

TBHQ:

This acronym stands for tertiary butylhydroquinone (t-butylhydroquinone), another chemistry mouthful. This synthetic, white to tan crystalline substance with a slight odor is an anti-oxidant that prevents oils and fats from going rancid and causing food to develop off-tastes and colors. However, of all the preservatives discussed so far, this additive is the most controversial.

Although the European Food Safety Authority (EFSA) and the FDA consider TBHQ to be noncarcinogenic, the Center for Science in the Public Interest recommends avoiding it. Rats fed TBHQ have exhibited a high incidence of tumors. However, the verdict is still out whether those findings translate to humans. Also, the additive has been shown to cause potential immunotoxin effects. The National Institute of Health (NIH) has observed neurological symptoms related to TBHQ consumption such as vision disturbances, convulsions, and breathing problems.

Because TBHQ is ubiquitous in the US food supply, studies have shown that 90% of Americans consume enough of it to meet or exceed the Acceptable Daily Intake (ADI) of 0.7mg per kg of body weight. For a 150-lb person, the ADI would be 48mg of TBHQ. At the FDA upper limit of 200mg per kg of refined oil, that person would need to consume a little over a half-pound of oil containing TBHQ every day to cause health problems. That's a bunch of oil. However, in frozen fish the levels of TBHQ can be as high as 1000 mg/kg. Also, TBHQ may be bioaccumulative, which would increase exposure over time. So, for consumers of high-fat diets and frozen fish, they may be at risk for TBHQ toxicity.

Given that TBHQ is an additive with a suspicious health rap, it is probably best to avoid eating foods that contain it, particularly since there are alternative anti-oxidants, like tocopherols, which are much safer.

Lastly, notice how many preservatives are in the Cinnabon® Sticks. Here's the summary: Calcium Disodium EDTA, sodium benzoate, potassium sorbate, TBHQ, and ascorbic acid (discussed for an earlier food item). A great deal of food chemistry went into the design of this dessert to keep it from spoiling over time.

Further Readings:
USA Today, *Pop-Tarts, Rice Krispies Treats, Cheez-Its Contain Preservative That May Harm Immune System, Study Says*

https://tinyurl.com/232c37rg

What About a Whole Pizza Hut Meal?

All of the food and beverage examinations so far have dealt with individual menu items. Let's take a quick look at a possible Pizza Hut meal: Meat Lover's® Original Stuffed Crust® (2 slices), Zesty Italian Salad, and Cinnabon Sticks (5 pieces).

MEAT LOVER'S® ORIGINAL STUFFED CRUST®:

APPLEWOOD SMOKED BACON: Bacon(1) (Cured with Water(2), Salt(3), Sugar(4), Sodium Phosphates(5), Sodium Erythorbate(6), Sodium Nitrite(7)), May Contain: Smoke Flavoring(8), Brown Sugar(9).

SLOW-ROASTED HAM: Ham(10) (Cured with Water(11), Salt(12), Sodium Lactate(13), Sugar(14), Sodium Phosphates(15), Sodium Diacetate(16), Sodium Erythorbate(17), Sodium Nitrite(18).

BEEF: Beef(19), Water(20), Salt(21), Tomato Paste(22), Natural Flavors(23), Dextrose(24), Dehydrated Onion(25), Spice(26).

ITALIAN SAUSAGE Pork(27), Seasoning (Spices (Including Mustard Seed(28), Salt(29), Paprika(30), Garlic Powder(31), Sugar(32), Natural Flavors(33), Extractives of Paprika(34)), Water(35).

PEPPERONI: Pork(36), Beef(37), Salt(38), Contains 2% or less of: Spices(39), Dextrose(40), Lactic Acid Starter Culture(41), Natural Spice Extractives(42), Extractives of Paprika(43), Extractive of Rosemary(44), Sodium Nitrite(45).

SEASONED PORK: Pork(46), Water(47), Salt(48), Spices(49), Sugar(50), Corn Syrup Solids(51), Autolyzed Yeast Extract(52), Natural Flavors(53).

CLASSIC MARINARA: Tomato Puree (Water(54), Tomato Paste(55)), Maltodextrin(56), Contains 2% or less of the following: Salt(57), Spices(58), Garlic Powder(59), Tomato Fibers(60), Olive Oil(61), Canola Oil(62), Citric Acid(63), Natural Flavors(64).

CHEESE: Mozzarella Cheese (Pasteurized Milk(65), Cheese Cultures(66), Salt(67), Enzymes(68)), Modified Food Starch(69), Sugar Cane Fiber(70), Potassium Chloride(71), Natural Flavors(72), Ascorbic Acid(73) To Protect Flavor.

ORIGINAL STUFFED CRUST: Enriched Flour (Bleached Wheat Flour(74), Malted Barley Flour(75), Niacin(76), Ferrous Sulfate(77), Thiamine Mononitrate(78), Riboflavin(79), Folic Acid(80)), Water(81), Yeast(82), Soybean Oil (83) Contains 2% or less of: Salt(84), Vital Wheat Gluten(85), DATEM(86), Sugar(87), Enzymes(88), Ascorbic Acid(89), Sucralose(90), Soybean Oil(91), TBHQ(92), Mozzarella Cheese (Pasteurized Milk(93), Cheese Cultures(94), Salt(95), Potassium Chloride(96), Enzymes(97)), Modified Food Starch(98), Nonfat Milk(99), Flavors(100).

Buttery Blend Crust Flavor (Medium or Large Pizza): Soybean Oil(101), Water(102), Salt(103), Vegetable Mono & Diglycerides(104), Contains Less Than 2% Of: Sodium Benzoate(105) and Potassium Sorbate(106) (As Preservatives), Soy Lecithin(107), Citric Acid(108), Natural Flavor(109), Calcium Disodium EDTA(110) Added To Protect Flavor, Beta Carotene(111) (Color), Vitamin A Palmitate(112) Added.

ZESTY ITALIAN SALAD:
SEASONED CROUTONS:

Enriched Flour (Wheat Flour(113), Malted Barley Flour(114), Niacin(115), Reduced Iron(116), Thiamine Mononitrate(117), Riboflavin(118), Folic Acid(119)), Canola(120) and/or Sunflower Oil(121), Whey(122), Salt(123), Yeast(124), 2% or Less of: High Fructose Corn Syrup(125), Sugar(126), Onion Powder(127), Dehydrated Parsley(128), Calcium Propionate(129) (Preservative), Calcium Peroxide(130), Calcium Sulfate(131), Ascorbic Acid(132), Azodicarbonamide(133), Enzymes(134), Sodium Stearoyl Lactylate(135), Spice Extractive(136), Spices(137), Paprika(138) (Color), Turmeric(139) (Color), Extractive of Paprika(140) (Color),

Citric Acid(141), TBHQ(142) (To Preserve Freshness). **PARMESAN CHEESE:** Pasteurized Milk(143), Cheese Cultures(144), Salt(145), Enzymes(146)), Powdered Cellulose(147) (Anticaking Agent). **LEAFY GREENS BLEND:** Iceberg Lettuce(148), Green Leaf Lettuce(149). **FRESH RED ONIONS:** Red Onions(150). **MEDITERRANEAN BLACK OLIVES:** Ripe Olive Wedges(151), Water(152), Salt(153), Ferrous Gluconate(154). **SLICED BANANA PEPPERS:** Fresh Banana Peppers(155), Vinegar(156), Water(157), Salt(158), Contains Less Than 2% of Lactic(159) and Malic Acid(160), Calcium Chloride(161), Sodium Benzoate(162), Polysorbate 80(163), Sodium Metabisulfite(164), Turmeric Extract(165). **SLICED ROMA TOMATOES:** Roma Tomatoes(166). **SLOW-ROASTED HAM:** Ham(167), Cured With: Water(168), Salt(169), Sodium Lactate(170), Sugar(171), Sodium Phosphates(172), Sodium Diacetate(173), Sodium Erythorbate(174), Sodium Nitrite(175).

PEPPERONI: Pork(176), Beef(177), Salt(178), Contains 2% or Less Of: Spices(179), Dextrose(180), Lactic Acid Starter Culture(181), Natural Spice Extractives(182), Extractives of Paprika(183), Extractives of Rosemary(184), Sodium Nitrite(185).

CINNABON STICKS:

Enriched Flour Bleached (Wheat Flour(186), Malted Barley Flour(187), Niacin(188), Ferrous Sulfate(189), Thiamine Mononitrate(190), Riboflavin(191), Folic Acid(192)), Water(193), Yeast(194), Salt(195), Soybean Oil(196), High Fructose Corn Syrup(197), DATEM(198), Vital Wheat Gluten(199), Enzymes(200), Ascorbic Acid(201), Sodium Stearoyl Lactylate(202). Pan Oil: Soybean Oil(203), TBHQ(204) Added To Protect Freshness. Buttery Blend: Soybean Oil(205), Water(206), Salt(207), Vegetable Mono & Diglycerides(208), Contains Less Than 2% Of: Sodium Benzoate(209) And Potassium Sorbate(210) (As Preservatives), Soy Lecithin(211), Natural Flavor(212), Citric Acid(213), Calcium

Disodium EDTA(214) Added To Protect Flavor, Beta Carotene(215) (Color), Vitamin A Palmitate(216) Added. **TOPPING:** Cinnamon(217) and Sugar(218).

It somewhat boggles the mind to observe that this meal, even without an added beverage, contains 218 total ingredients. This observation is typical of ultra-processed foods: foods, engineered by food scientists, that require a host of synthetic chemicals to add flavor, color, texture, protection from spoilage, protection from unfavorable changes over time, hold incompatible components together like oil and water, alter and fix acidity levels, and replace nutrients that were removed from the natural food during processing.

The ingredients highlighted in gray have been deemed significantly processed. There are 138 of them in this meal. That's 63.3% of the total or, another way of putting it, about 6 out of 10 of the ingredients in this meal are significantly processed. That's another characteristic of ultra-processed foods: they are exceptionally refined or synthesized in factories causing them to be only remotely related to foods in their natural state.

What about the caloric content of this meal? Consuming two pieces of pizza contributes 860 calories. The salad has 360 calories. Five pieces of the cinnamon pastry provides 400 calories. The sum total for the meal is 1620 calories. If 2000 calories per day is the target for the average adult American, the calorie content of this meal amounts to 81% of the recommended daily amount. That only leaves 380 calories for the remaining meals and snacks of the day. And that's another characteristic of ultra-processed, high caloric-density foods: the calories add up fast making it difficult for a person to maintain a healthy weight and body mass index (BMI).

Look at the ingredients in this meal, but this time pay attention to the vitamins. They show up in multiple places. Whenever wheat shows up in fast food, and that's frequently, the consumer gets a shot of vitamins due to enrichment, a topic discussed earlier. Can a person overdose on added vitamins sourced from processed foods? Maybe. It's certainly a possibility for children who have lower daily requirements than adults. According to Mother Jones Magazine, "Studies have shown a host of illnesses associated with excessive intake of these nutrients (vitamin A, zinc, and niacin) ... according to the EWG (Environmental Working Group)."

Further Readings:
Mother Jones Magazine, "*Is Your Cereal Giving You a Vitamin Overdose?*"

https://tinyurl.com/272n37n5

CHAPTER 13

INGREDIENTS IN SELECTED TACO BELL MENU ITEMS

"The fresh food place"

— Taco Bell slogan

General Comments

In my younger days as a fast-food junkie, probably my all-time favorite drive-through restaurant was Taco Bell. Yes, the menu was full of highly Americanized Mexican food and not very authentic, but I liked all the options and combinations, particularly the soft burritos. Nowadays, maybe once a year, I'll trot down memory lane and place an order for those burritos accompanied by hot sauce condiments. That choice feels good for a few hours, but, by the next day, I've had buyer's remorse.

Of course, the Taco Bell menu is full of Mexican-sounding names like quesarito, burrito, chalupa, quesalupa, nachos, and quesadilla. Of course, what it all comes down to is a combination of meat, cheese, vegetables, and sauces in a seemingly endless variety served up in soft shells, hard shells, on a bed of Spanish rice, wraps, chips, or other floured carriers. There are a few specialty beverages offered, but, oddly, there are slim pickings for desserts, even though they are popular in other Mexican restaurants.

Taco Bell began as its own entity in 1948 with a store opened in San Bernardino, California, called Bell's Drive-in, named after the owner, Glen Bell (yes, his name was actually "Bell.") It didn't take on the Mexican theme until 1951, when its name was changed to Taco-Tia, then El Taco, and finally Taco Bell. Taco Bell's independence lasted until 1978 when it was purchased by PepsiCo. By the early 2000s, Taco Bell merged with Yum! Brands, which managed other fast-food restaurants like A&W Restaurants, Long John Silver's, and later Pizza Hut and Kentucky Fried Chicken.

As a reminder, all of the surveyed Taco Bell menu items, including the ones mentioned below, can be viewed at: https://tinyurl.com/4fhefvxs

The Taco Bell menu items selected for in-depth looks are:

- **Power Menu Bowl**

- **Quesarita - Steak**

- **Grande Nachos - Steak**

- **Wild Strawberry Freeze™**

Taco Bell's Power Menu Bowl – Steak (480 Calories)

Ingredients:

Avocado Ranch Sauce: Soybean Oil(1), Cultured Buttermilk(2), Water(3), Avocado(4), Cage-Free Egg Yolk(5), Vinegar(6), White Wine Vinegar(7), 2% or less of: Salt(8), Garlic Juice(9), Sugar(10), Natural Flavor(11), Lemon(12) and Lime Juice Concentrate(13), Spice(14), Garlic Powder(15), Onion Powder(16), Onion(17), Garlic(18), Lactic Acid(19), Citric Acid(20), Ascorbic Acid(21), Turmeric Oleoresin(22), Fruit Juice(23), Fruit Juice Concentrate(24), Disodium Inosinate(25) and Disodium Guanylate(26), Propylene Glycol Alginate(27), Xanthan Gum(28), Sodium Alginate(29), Potassium Sorbate(30), Sodium Benzoate(31), Calcium Disodium EDTA(32). **Cheddar Cheese**: Cheddar Cheese (Cultured

Pasteurized Milk(33), Salt(34), Enzymes(35), Annatto(36), Anti-Caking Agent (X).

Reduced-Fat Sour Cream: Milk(37), Cream(38), Modified Corn Starch(39), Lactic Acid(40), Maltodextrin(41), Citric Acid(42), Sodium Phosphate(43), Natural Flavor(44), Cellulose Gel(45), Potassium Sorbate(46), Cellulose Gum(47), Guar Gum(48), Locust Bean Gum(49), Carrageenan(50), Vitamin A(51).

Seasoned Rice: Enriched Long Grain Rice(52), Water(53), Canola Oil(54), Seasoning (Maltodextrin(55), Salt(56), Natural Flavors(57), Tomato Powder(58), Sugar(59), Garlic Powder(60), Spices(61), Onion(62), Tomato(63), Red(64) and Green Bell Peppers(65), Citric Acid(66), Paprika(67), Onion Powder(68), Paprika(69), Disodium Guanylate(70) and Inosinate(71), Torula Yeast(72).

Black Beans: Black Beans(73), Water(74), Onion(75), Canola Oil(76), Seasoning (Water(77), Dextrose(78), Salt(79), Natural Flavor(80), Corn Starch(81), Corn Oil(82), Onion Powder(83), Garlic Powder(84), Turmeric(85)), Modified Corn Starch(86), Salt(87), Chili Powder(88), Garlic(89), Onion(90).

Guacamole: Avocado(91), Water(92), Tomato(93), Onion(94), Jalapeno(95), Salt(96), Cilantro(97), Lemon(98) Or Lime Juice(99), Ascorbic(100) or Erythorbic Acid(101), Xanthan Gum(102), Sodium Alginate(103).

USDA Select Marinated Grilled Steak: Beef(104), Water(105), Seasoning (Modified Potato Starch(106), Natural Flavors(107), Salt(108), Brown Sugar(109), Dextrose(110), Carrageenan(111), Dried Beef Stock(112), Cocoa Powder(113), Onion Powder(114), Disodium Inosinate(115) & Guanylate(116), Tomato Powder(117), Corn Syrup Solids(118), Maltodextrin(119), Garlic Powder(120), Spice(121), Citric Acid(122), Lemon Juice Powder(123)), Sodium Phosphates(124). Sauce: Water(125), Seasoning (Natural Flavors(126), Dextrose(127), Brown Sugar(128), Salt(129), Dried Beef Stock(130), Onion Powder(131), Tomato Powder(132), Corn

Syrup Solids(133), Maltodextrin(134), Disodium Inosinate(135) & Guanylate(136), Garlic Powder(137), Spices(138), Cocoa Powder(139), Citric Acid(140), Lemon Juice Powder(141)).

Ingredient Assignments and the Degree of Industrialization (Processing)

Annatto (36) Moderately; **Ascorbic Acid** (synthetic) (21,100) Extremely; **Brown Sugar** (109,128) Moderately; **Calcium Disodium EDTA** (synthetic) (32) Extremely; **Canola Oil** (54,76) Highly; **Carrageenan** (50,111) Moderately; **Cellulose Gel** (45) Extremely; **Cellulose Gum** (synthetic) (47) Extremely; **Citric Acid** (synthetic) (20,42,66,122,140) Extremely; **Corn Oil** (82) Highly; **Corn Starch** (81) Highly; **Corn Syrup Solids** (synthetic) (118,133) Extremely; **Dextrose** (synthetic) (78,110,127) Extremely; **Disodium Guanylate** (synthetic) (26,70,116,136) Extremely; **Disodium Inosinate** (synthetic) (25,71,115,135) Extremely; **Enzymes** (35) Highly; **Erythorbic Acid** (synthetic) (101) Extremely; **Guar Gum** (48) Moderately; **Lactic Acid** (synthetic) (19,40) Extremely; **Locust Bean Gum** (49) Highly; **Maltodextrin** (41,55,119,134) Highly; **Modified Corn Starch** (39,86) Extremely; **Modified Potato Starch** (106) Extremely; **Natural Flavors** (11,44,80,107,126) Moderately; **Potassium Sorbate** (synthetic) (30,46) Extremely; **Propylene Glycol Alginate** (synthetic) (27) Extremely; **Sodium Alginate** (29,103) Extremely; **Sodium Benzoate** (synthetic) (31) Extremely; **Sodium Phosphate** (synthetic) (43,124) Extremely; **Soybean Oil** (1) Highly; **Sugar** (10,59) Moderately; **Turmeric Oleoresin** (22) Highly; **Vinegar** (6) Moderately; **Vitamin A** (synthetic) (51) Extremely; **White Wine Vinegar** (7) Moderately; **Xanthan Gum** (synthetic) (28,102) Extremely.

Reporting the Processed Food Index (PFI) for the Power Menu Bowl

# OF UNPROCESSED INGREDIENTS:	16
# OF LIGHTLY PROCESSED INGREDIENTS:	23
# OF MODERATELY PROCESSED INGREDIENTS:	8
# OF HIGHLY PROCESSED INGREDIENTS:	8
# OF EXTREMELY PROCESSED INGREDIENTS:	20

TOTAL # OF INGREDIENTS:	141
# OF UNIQUE INGREDIENTS:	75 (53%)
PROCESSED FOOD INDEX:	34 (Moderately Industrialized)

Commentary

This menu item is a bit of an enigma. There are more ingredients in this dish than in all the menu items discussed so far: an astounding 141 ingredients. However, the Processed Food Index is 34 (moderately industrialized), which is one of the lowest scores for a menu dish. What gives? The big clue is the number of unprocessed and lightly processed ingredients, i.e., 39, or 52% of the unique ingredients. Looking at the ingredient list, you see spices, vegetables, beef, milk, water, and lightly processed components like garlic powder, onion powder, tomato powder, and cocoa powder. There are only 20 extremely processed ingredients, or 27%. In terms of a healthier choice for fast food, the Power Menu Bowl is a much better option compared to most other foods discussed so far. Also, the calorie count is only 480, pretty low for a main dish. If you choose to consume fast food, look for menu items with low PFIs … the lower the better. Unfortunately, in the current year, 2022, a PFI score or something equivalent for fast food does not exist when you order menu items, making it difficult for consumers to make intelligent choices. The only data consumers have available to them are calorie counts and nutrition facts, if they choose to seek out this information at the restaurant facility or the company's website.

The highlighted ingredients in Taco Bell's Power Menu Bowl are:

- Carrageenan
- Citric Acid
- Enzymes
- Propylene Glycol Alginate (PGA)

Carrageenan:

This additive is found in the reduced fat sour cream and the marinated grilled steak. Carrageenan is a very common additive found in many processed foods. It's typically used as an emulsifier, stabilizer, and thickener. In the sour cream portion of the Power Menu Bowl, most likely the carrageenan serves to replace fat and provide creaminess. In meat products like the grilled steak, it serves as a binder since it strongly clings to proteins.

Carrageenan is a natural substance found in seaweed. The most popular source for food products is red edible seaweed, commonly known as Irish moss: a red, parsley-like plant. In the West, its use in foods can be traced back to the 1400s. Resistant to digestion, this additive is considered a beneficial dietary fiber.

There are several ways to manufacture carrageenan for food use. Here is one way: The harvested Irish moss is cleaned and washed to remove salt and sand. Then it is cooked in hot alkali to extract the carrageenan. The extract is filtered to remove insoluble residue. Alcohol is added to the extract to separate the solid carrageenan, which is then dried and milled to produce refined carrageenan.

Depending upon treatment during manufacturing, two different carrageenan products can be made. If acid is used instead of alkali to extract carrageenan from the seaweed, then a product called "degraded carrageenan" or poligeenan is produced.

Not only is carrageenan one of the most common food additives, but it's also one of the most controversial. Both the FDA and the European Food Safety Authority have concluded that it's safe for human consumption at consumption levels of 34mg per pound of body weight per day. For a 150-lb person, that equates to 5.1g, which is a significant amount. However, some scientific studies, as well as anecdotal reports, indicate some serious health issues, such as ulcerative colitis, fetal toxicity and birth defects, colorectal and liver cancer, insulin resistance, inflammation, and immune suppression. That's quite a list but those studies rely on the testing of lab animals, like mice, and there is no direct evidence that carrageenan is unsafe for humans. Some researchers believe that the poligeenan form is responsible for the negative health effects, whereas the alkali-derived form is absolutely safe at recommended levels. Today, the controversy rages on in the scientific community. However, the National Organic Standards Board voted to remove carrageenan as an approved additive for organic foods. If consumers have any concern at all about the health safety of carrageenan, they should just avoid consumption of foods containing it. Of course, that's a problem for fast-food consumers, since they are unlikely to know when carrageenan is present. Fortunately, for commercial processed foods available in grocery stores and food marts, there are numerous other stabilizers and thickeners that can replace carrageenan such as agar-agar, guar gum, gelatin, and pectin.

Further Readings:
Food Crumbles, *"The Science of Carrageenan"*
https://foodcrumbles.com/science-carrageenan-thickening-gelling-foods-seaweed/

Dr. Axe, *"Is Carrageenan Bad for Your Health?"*

https://draxe.com/nutrition/what-is-carrageenan/

Citric Acid:

This is a ubiquitous additive in the fast-food industry with applications in other industries as well. It can be found in citrus fruits like lemons and limes at up to 8% of the dry weight. Although a natural substance and produced in the human body, citric acid is usually manufactured in a chemical plant using complicated, industrial processes and hazardous chemicals (see below). Here are some of its uses: (1) in cheese sauces to provide tartness, (2) in baked goods to create a slightly acidic environment to promote fermentation in dough, (3) in meats to assist with curing, (4) in ice cream as an emulsifying agent to keep fats from separating, (5) in soft drinks as a flavoring agent and preservative, and (6) to replace lemon juice in recipes. As an acidifying agent, it can help remove off-flavors.

In 1917, it was discovered that certain strains of the mold aspergillus niger (A. niger) could efficiently make citric acid through a fermentation process. Although other methods of making citric acid exist, the fermentation process is today the leading industrial method. The mold organisms, which may be genetically modified to bolster output, are fed simple sugars, like glucose, in a fermentation vessel and incubated for a period of time until maximum production of citric acid is reached. There are several industrial methods to isolate the citric acid. One of the major methods involves solvent extraction. In this process, the aqueous fermentation broth is combined with a mixture of organic liquids, namely the solvent n-octyl alcohol (a natural substance) and the extractant tridodecyl amine (an eye, skin, and lung irritant). The organic phase is mechanically separated from the aqueous one. The crystallized citric acid is separated from the organic phase by treatment with water, after which it is purified.

Enzymes:

This is a category of chemicals and one of the most generic ingredients listed in processed foods. When this ingredient is listed for a product, there is really no useful information provided to the consumer. The FDA allows such generic usage to protect proprietary interests in the food industry.

Enzymes are protein molecules. They are biological catalysts that greatly speed up biochemical reactions, e.g., amylases that break down cornstarch into simple sugars (corn syrup). Note that all enzyme names end in "-ase" designating the molecules as enzymes. There are literally thousands and thousands of enzymes that drive huge numbers of reactions. In the food industry, isolated enzymes, extracted from edible plants and the tissues of animals, are used in the preparation of beer, wine, cheese, and bread. Also, microorganisms like bacteria, yeasts, and fungi are sources of enzymes. For example, rennet is an enzyme mixture isolated from the stomachs of calves and used to curdle cheese.

In the Power Menu Bowl, enzymes are listed as ingredients for the cheese. Since enzymes are natural chemicals, they are considered safe additives.

Propylene Glycol Alginate (PGA):

This synthetic additive is found in the avocado ranch sauce. It's a white, water-soluble powder that has multiple uses in the food industry but mainly serves as a stabilizer, thickener, and emulsifier. It's synthesized from alginic acid, found in the cell walls of brown algae, and the synthetic chemical propylene glycol.

PGA thickens liquids thereby stabilizing them. It's used as a beer foam stabilizer to maintain the integrity of the foam. Most types of gel-like foods contain PGA, such as yogurt, jelly, ice cream, and dressings. Although it is considered a safe additive, there have been reports of negative side effects, such as stomach upset, nausea, and allergic reactions,

Taco Bell's Quesarito – Steak (630 Calories)

Ingredients:

Flour Tortilla: Enriched Wheat Flour(1), Water(2), Vegetable Shortening (Soybean(3), Hydrogenated Soybean(4) and/or Cottonseed Oil(5)), Sugar(6), Salt(7), Leavening (Baking Soda(8), Sodium Acid Pyrophosphate(9)), Molasses(10), Dough Conditioner (Fumaric Acid(11), Distilled Monoglycerides(12), Enzymes(13), Vital Wheat Gluten(14), Cellulose Gum(15), Wheat Starch(16), Calcium Carbonate(17)), Calcium Propionate(18), Sorbic Acid(19), and/or Potassium Sorbate(20).

Cheddar Cheese: Cheddar Cheese (Cultured Pasteurized Milk(21), Salt(22), Enzymes(23), Annatto(24), Anti-Caking Agent (X).

Reduced-Fat Sour Cream: Milk(25), Cream(26), Modified Corn Starch(27), Lactic Acid(28), Maltodextrin(29), Citric Acid(30), Sodium Phosphate(31), Natural Flavor(32), Cellulose Gel(33), Potassium Sorbate(34), Cellulose Gum(35), Guar Gum(36), Locust Bean Gum(37), Carrageenan(38), Vitamin A(39).

Nacho Cheese Sauce: Nonfat Milk(40), Cheese Whey(41), Water(42), Vegetable Oil (Canola Oil(43), Soybean Oil(44)), Modified Food Starch(45), Maltodextrin(46), Natural Flavors(47), Salt(48), Dipotassium Phosphate(49), Jalapeno Puree(50), Vinegar(51), Lactic Acid(52), Cellulose Gum(53), Potassium Citrate(54), Sodium Stearoyl Lactylate(55), Citric Acid(56), Annatto(57) and Oleoresin Paprika(58).

Seasoned Rice: Enriched Long Grain Rice(59), Water(60), Canola Oil(61), Seasoning (Maltodextrin(62), Salt(63), Natural Flavors(64), Tomato Powder(65), Sugar(66), Garlic Powder(67), Spices(68), Onion(69), Tomato(70), Red(71) And Green Bell Peppers(72),

Citric Acid(73), Paprika(74), Onion Powder(75), Paprika(76), Disodium Guanylate(77) and Inosinate(78), Torula Yeast(79).

Creamy Chipotle Sauce: Soybean Oil(80), Water(81), Vinegar(82), Cage-Free Egg Yolk(83), 2% or less of Chili Peppers(84), Chipotle Peppers(85), Salt(86), Sugar(87), Roasted Garlic(88), Natural Flavor(89), Natural Smoke Flavor(90), Garlic Powder(91), Onion Powder(92), Paprika Extract(93), Xanthan Gum(94), Propylene Glycol Alginate(95), Sodium Benzoate(96), Potassium Sorbate(97), Calcium Disodium EDTA(98), Citric Acid(99), Potassium Chloride(100), Maltodextrin(101).

USDA Select Marinated Grilled Steak: Beef(102), Water(103), Seasoning (Modified Potato Starch(104), Natural Flavors(105), Salt(106), Brown Sugar(107), Dextrose(108), Carrageenan(109), Dried Beef Stock(110), Cocoa Powder(111), Onion Powder(112), Disodium Inosinate(113) & Guanylate(114), Tomato Powder(115), Corn Syrup Solids(116), Maltodextrin(117), Garlic Powder(118), Spice(119), Citric Acid(120), Lemon Juice Powder(121)), Sodium Phosphates(122). Sauce: Water(123), Seasoning (Natural Flavors(124), Dextrose(125), Brown Sugar(126), Salt(127), Dried Beef Stock(128), Onion Powder(129), Tomato Powder(130), Corn Syrup Solids(131), Maltodextrin(132), Disodium Inosinate(133) & Guanylate(134), Garlic Powder(135), Spices(136), Cocoa Powder(137), Citric Acid(138), Lemon Juice Powder(139)).

Ingredient Assignments and the Degree of Industrialization (Processing)

Annatto (24) Moderately; **Baking Soda** (synthetic) (8) Extremely; **Brown Sugar** (107,126) Moderately; **Calcium Carbonate** (synthetic) (17) Extremely; **Calcium Disodium EDTA** (synthetic) (98) Extremely; **Calcium Propionate** (synthetic) (18) Extremely; **Canola Oil** (43,61) Highly; **Carrageenan** (38,109) Moderately; **Cellulose Gel** (33) Extremely; **Cellulose Gum** (synthetic) (15,35,53)Extremely; **Cheese Whey** (41) Moderately; **Citric Acid**

(synthetic) (30,56,73,99,120,138) Extremely; **Corn Syrup Solids** (synthetic) (116,131) Extremely; **Cottonseed Oil** (5) Highly; **Dextrose** (synthetic) (108,125) Extremely; **Dipotassium Phosphate** (synthetic) (49) Extremely; **Disodium Guanylate** (synthetic) (77,114,134) Extremely; **Disodium Inosinate** (synthetic) (78,113,133) Extremely; **Distilled Monoglycerides** (synthetic) (12) Extremely; **Enzymes** (13,23) Highly; **Fumaric Acid** (synthetic) (11) Extremely; **Guar Gum** (36) Moderately; **Hydrogenated Soybean Oil** (synthetic) (4) Extremely; **Lactic Acid** (synthetic) (28,52) Extremely; **Locust Bean Gum** (37) Highly; **Maltodextrin** (29,46,62,101,117,132) Highly; **Modified Corn Starch** (27) Extremely; **Modified Food Starch** (45) Extremely; **Modified Potato Starch** (104) Extremely; **Natural Flavors** (32,47,64,89,105,124) Moderately; **Oleoresin Paprika** (58) Highly; **Paprika Extract** (93) Moderately; **Potassium Chloride** (synthetic) (100) Extremely; **Potassium Citrate** (synthetic) (54) Extremely; **Potassium Sorbate** (synthetic) (20,34,97) Extremely; **Propylene Glycol Alginate** (synthetic) (95) Extremely; **Smoke Flavor** (90) Highly; **Sodium Benzoate** (synthetic) (96) Extremely; **Sodium Phosphate** (synthetic) (31,122) Extremely; **Sodium Stearoyl Lactylate** (synthetic) (55) Extremely; **Sodium Acid Pyrophosphate** (synthetic) (9) Extremely; **Sorbic Acid** (synthetic) (19) Extremely; **Soybean Oil** (3,44,80) Highly; **Sugar** (6,66,87) Moderately; **Vinegar** (51,82) Moderately; **Vital Wheat Gluten** (14) Moderately; **Vitamin A** (synthetic) (39) Extremely; **Wheat Flour** (1) Moderately; **Wheat Starch** (16) Highly; **Xanthan Gum** (synthetic) (94) Extremely.

Reporting the Processed Food Index (PFI) for Steak Quesarito

# OF UNPROCESSED INGREDIENTS:	13
# OF LIGHTLY PROCESSED INGREDIENTS:	14
# OF MODERATELY PROCESSED INGREDIENTS:	11
# OF HIGHLY PROCESSED INGREDIENTS:	9
# OF EXTREMELY PROCESSED INGREDIENTS:	30

TOTAL # OF INGREDIENTS:	139
# OF UNIQUE INGREDIENTS:	77 (55%)
PROCESSED FOOD INDEX:	47 (Highly Industrialized)

Commentary

This Quesarito is a pretty complex wrap. Its 7 components are: flour tortilla, cheddar cheese, reduced-fat sour cream, nacho cheese sauce, seasoned rice, creamy chipotle sauce, and marinated grilled steak. With 139 ingredients, this is another menu item packed with a host of foods and additives. Given that 77 of the 139 ingredients are unique, no one could possibly re-create the quesarito in their home kitchen. The reason is provided by the Processed Food Index of 47, indicating a highly industrialized product containing 39 highly and extremely processed ingredients. Also, with 630 calories, the quesarito is not a light menu item. If a consumer ate two of them (probably not unusual), they would net 1260 calories or roughly 63% of the daily recommendation.

The highlighted ingredients in Taco Bell's Quesarito - Steak are:

- Fumaric Acid
- Locust Bean Gum
- Sodium Acid Pyrophosphate (SAPP)
- Vital Wheat Gluten (VWG)

Fumaric Acid:

This ingredient is found in the flour tortilla and is listed as a dough conditioner. It's a white, crystalline material with a fruity taste. Fumaric acid is naturally occurring in the plant world where it is found in mushrooms, lichens, and Iceland moss. It's also formed in the human body during metabolic processes, and human skin can produce it when exposed to sunlight.

Fumaric acid in foods is synthetically made. Fumaric acid has a sister compound called maleic acid which is similar in structure. Maleic acid can be converted into fumaric acid using a reaction called catalytic isomerization.

Fumaric acid has a number of uses in processed foods. As an acidulant, it can increase the acidity (lower pH) of foods, thereby acting as a preservative. It can take the place of citric acid, and, since it is 1.5 times more acidic, less of it has to be used. That's probably why it is used in the flour tortilla since fumaric acid is also a common dough conditioner. A dough conditioner is any baking ingredient that improves the production and consistency of a dough. A dough conditioner simplifies and assists the bread-making process.

Fumaric acid is considered a very safe food additive.

Locust Bean Gum (Carob Bean Gum):

This substance is found in the reduced-fat sour cream. It's one of many vegetable gums that are used in processed foods. Gums are essentially thickeners which impart a gel-like consistency to foods. Locust bean gum is extracted from the seeds of the carob tree (or locust tree), which is grown in several countries in the Mediterranean region. Interestingly, a chocolate-like powder (carob) can be made from the fruit pod after the seeds are removed, and it has some properties that make it similar to cacao powder (chocolate). The locust bean gum is a natural off-white powder. Finally, locust bean gum is a good source of soluble fiber.

In the reduced-fat sour cream, the locust bean gum is accompanied by cellulose gum, guar gum, and carrageenan, the combination of which provides the texture and creaminess expected in conventional sour cream.

Locust bean gum is very safe as a food additive although some people may have an asthma-like reaction towards it.

Sodium Acid Pyrophosphate (SAPP or disodium dihydrogen pyrophosphate or disodium dihydrogen diphosphate):

This very chemical sounding additive is very popular in the processed food industry, particularly the baking industry, and has been around since the early 1900s. It is found in the flour tortilla. SAPP functions as a leavening agent, a chemical which reacts with one or two other chemicals to produce carbon dioxide gas causing dough to rise during baking. In looking at the ingredient list for the flour tortilla, SAPP sits next to baking soda (sodium bicarbonate), its counterpart in the leavening process. The leavening reaction is actually a neutralization where the base (sodium bicarbonate) reacts with the acid (SAPP) to produce carbon dioxide. In the home kitchen, SAPP is often found in double-acting baking powder and self-rising flour. SAPP is a synthetic white powder with a bit of an off taste, so it's often mixed with sweet foods to mask the taste of it. Notice that there is sugar in the flour tortilla. SAPP can also be found in potato products, like hash browns, where it keeps the potatoes from darkening. It also has applications as an emulsifier in cheeses and as an accelerator of the curing process for meats.

SAPP has a good health record, particularly since it has been used for over a century without serious issues.

Vital Wheat Gluten (VWG):

This is the natural, protein component of wheat flour and the boogie man for anybody with celiac disease, gluten intolerance, or gluten sensitivity. However, in the baking industry it plays a pivotal role by giving the baked product structure and assisting the dough in rising. Kneading bread dough promotes the formation of the gluten network creating products that are chewy. Bread flours are high in gluten versus pastry flours which are low.

As expected, the vital wheat gluten is found in the flour tortilla.

Vital wheat gluten is produced commercially by washing regular wheat flour until the starch component is completely removed. The playdough-like material that remains is then dried, pulverized, and sold as a powdered product. Another very popular use of vital wheat gluten is in Asian foods where it can be texturized to mimic meats, e.g., mock duck. In that form it is called seitan. With the exception of people who are gluten sensitive, VWG is safe to consume.

Taco Bell's Grande Nachos – Steak (1080 Calories)

Ingredients:

Nacho Cheese Sauce: Nonfat Milk(1), Cheese Whey(2), Water(3), Vegetable Oil (Canola Oil(4), Soybean Oil(5)), Modified Food Starch(6), Maltodextrin(7), Natural Flavors(8), Salt(9), Dipotassium Phosphate(10), Jalapeno Puree(11), Vinegar(12), Lactic Acid(13), Cellulose Gum(14), Potassium Citrate(15), Sodium Stearoyl Lactylate(16), Citric Acid(17), Annatto(18) and Oleoresin Paprika(19).

Nacho Chips: White Ground Corn(20), Water(21), 2% Or Less Of: Calcium Propionate(22), Fumaric Acid(23), Sorbic Acid(24),

Cellulose Gum(25), Sodium Propionate(26), Salt(27). Prepared in Canola(28) and/or Vegetable Oil(29).

Refried Beans: Pinto Beans(30), Soybean Oil(31), Seasoning (Salt(32), Sugar(33), Spice(34), Beet Powder(35), Natural Flavors(36), Sunflower Oil(37), Maltodextrin(38), Corn Flour(39), Trehalose(40), Modified Corn Starch(41)).

Guacamole: Avocado(42), Water(43), Tomato(44), Onion(45), Jalapeno(46), Salt(47), Cilantro(48), Lemon(49) Or Lime Juice(50), Ascorbic(51) or Erythorbic Acid(52), Xanthan Gum(53), Sodium Alginate(54).

Reduced-Fat Sour Cream: Milk(55), Cream(56), Modified Corn Starch(57), Lactic Acid(58), Maltodextrin(59), Citric Acid(60), Sodium Phosphate(61), Natural Flavor(62), Cellulose Gel(63), Potassium Sorbate(64), Cellulose Gum(65), Guar Gum(66), Locust Bean Gum(67), Carrageenan(68), Vitamin A(69).

Three Cheese Blend: Low-Moisture Part-Skim Mozzarella Cheese, Cheddar Cheese, Pasteurized Process Monterey Jack And American Cheese With Peppers (Cultured Milk(70), Cultured Part-Skim Milk(71), Water(72), Cream(73), Salt(74), Sodium Citrate(75), Jalapeno Peppers(76), Sodium Phosphate(77), Lactic Acid(78), Sorbic Acid(79), Color Added: Annatto(80) and Paprika Extract Blend(81), Enzymes(82), Anticaking Agents (Potato Starch(83), Cornstarch(84), Powdered Cellulose(85))).

Tomatoes: Fresh Tomatoes(86).

USDA Select Marinated Grilled Steak: Beef(87), Water(88), Seasoning (Modified Potato Starch(89), Natural Flavors(90), Salt(91), Brown Sugar(92), Dextrose(93), Carrageenan(94), Dried Beef Stock(95), Cocoa Powder(96), Onion Powder(97), Disodium Inosinate(98) & Guanylate(99), Tomato Powder(100), Corn Syrup Solids(101), Maltodextrin(102), Garlic Powder(103), Spice(104), Citric Acid(105), Lemon Juice Powder(106)), Sodium Phosphates(107). Sauce: Water(108), Seasoning (Natural

Flavors(109), Dextrose(110), Brown Sugar(111), Salt(112), Dried Beef Stock(113), Onion Powder(114), Tomato Powder(115), Corn Syrup Solids(116), Maltodextrin(117), Disodium Inosinate(118) & Guanylate(119), Garlic Powder(120), Spices(121), Cocoa Powder(122), Citric Acid(123), Lemon Juice Powder(124)).

Ingredient Assignments and the Degree of Industrialization (Processing)

Annatto (18,80) Moderately; **Ascorbic Acid** (synthetic) (51) Extremely; **Brown Sugar** (92,111) Moderately; **Calcium Propionate** (synthetic) (22) Extremely; **Canola Oil** (4,28) Highly; **Carrageenan** (68,94) Moderately; **Cellulose Gel** (synthetic) (63) Extremely; **Cellulose Gum** (synthetic) (14,25,65)Extremely; **Cheese Whey** (2) Moderately; Citric Acid (synthetic) (17,60,105,123) Extremely; Corn **Syrup Solids** (synthetic) (101,116) Extremely; **Cornstarch** (84) Highly; **Corn Flour** (39) Moderately; **Dextrose** (synthetic) (93,110) Extremely; **Dipotassium Phosphate** (synthetic) (10) Extremely; **Disodium Guanylate** (synthetic) (99,119) Extremely; **Disodium Inosinate** (synthetic) (98,118) Extremely; **Enzymes** (82) Highly; **Erythorbic Acid** (synthetic) (52) Extremely; **Fumaric Acid** (synthetic) (23) Extremely; **Guar Gum** (66) Moderately; Lactic Acid (synthetic) (13,58,78) Extremely; **Locust Bean Gum** (67) Highly; **Maltodextrin** (7,38,59,102,117) Highly; **Modified Corn Starch** (41,57) Extremely; **Modified Food Starch** (6) Extremely; **Modified Potato Starch** (89); Extremely; **Natural Flavors** (8,36,62,90,109) Moderately; **Oleoresin Paprika** (19) Highly; **Paprika Extract Blend** (81) Moderately; **Potassium Citrate** (synthetic) (15); Extremely; **Potassium Sorbate** (synthetic) (64) Extremely; **Potato Starch** (83) Moderately; **Powdered Cellulose** (85) Highly; **Sodium Alginate** (synthetic) (54) Extremely; **Sodium Citrate** (synthetic) (75) Extremely; **Sodium Stearoyl Lactylate** (synthetic) (16) Extremely; **Sodium Phosphates** (synthetic) (61,77,107) Extremely; **Sodium Propionate** (synthetic)

(26) Extremely; **Sorbic Acid** (synthetic) (24,79) Extremely; **Soybean Oil** (5,31) Highly; **Sugar** (33) Moderately; **Sunflower Oil** (37) Highly; **Trehalose** (40) Highly; **Vegetable Oil** (29) Highly; **Vinegar** (12) Moderately; **Vitamin A** (synthetic) (69) Extremely; **Xanthan Gum** (synthetic) (53) Extremely.

Reporting the Processed Food Index (PFI) for Steak Grande Nachos

# OF UNPROCESSED INGREDIENTS:	12
# OF LIGHTLY PROCESSED INGREDIENTS:	14
# OF MODERATELY PROCESSED INGREDIENTS:	11
# OF HIGHLY PROCESSED INGREDIENTS:	11
# OF EXTREMELY PROCESSED INGREDIENTS:	26

TOTAL # OF INGREDIENTS:	124
# OF UNIQUE INGREDIENTS:	74 (60%)
PROCESSED FOOD INDEX:	45 (Highly Industrialized)

Commentary

This nacho dish is even more complex that the Quesarito reviewed earlier. It contains the following 8 components: nacho cheese sauce, nacho chips, refried beans, guacamole, reduced-fat sour cream, three cheese blend, tomatoes, and marinated grilled steak. At 124, it doesn't skimp on the number of ingredients. This highly industrialized product has 48 significantly processed ingredients out of a possible 74 or 65%. But the most astounding feature of this dish is its calorie count: 1080. That's over half of the daily recommended amount for the average American female. The high calorie count is partly accounted for by vegetable oils showing up 4 times, cream 2 times, and the beef (at least 28% fat).

The highlighted ingredients in Taco Bell's Grande Nachos - Steak are:

- Calcium Propionate
- Sodium Alginate (Algin)
- Trehalose

Calcium Propionate:

This synthetic, white powder keeps baked goods from going moldy and, thus, acts as a mold inhibitor and preservative (extends shelf life). The anti-mold property of calcium propionate was discovered in the early 1900s, and it has been used as an additive in the baking industry since the 1930s. It's an ideal preservative since it doesn't adversely affect yeast or fermentation in the bread-making process. It also has applications in processed meat, whey, cheese, and other dairy products. It is found naturally in Swiss cheese at levels up to 1%. This additive also has anti-bacterial properties.

In the nacho dish, it's no surprise that calcium propionate shows up in the nachos along with its cousin, sodium propionate, another preservative.

Although there have been anecdotal reports about adverse effects, like digestive problems and migraine headaches, there haven't been any research studies showing harmful effects. The European Food Safety Authority in 2014 did not find any safety concerns for calcium propionate.

Sodium Alginate (Algin):

This additive has a natural source. It's derived from alginic acid found in brown seaweed (kelp is an example) harvested from the northern Atlantic Ocean. In combination with a source of calcium, e.g., calcium chloride, it can form a gel, so it acts as thickener or stabilizer in a food system. Sodium alginate is a white to yellowish-brown powder and is considered very safe as a food additive.

The additive is made from seaweed in the following process: (1) Seaweed is washed with water to remove soluble impurities; (2) Sodium carbonate is added to convert all algin compounds to a soluble form; (3) Cellulosic impurities are removed by filtration; (4) Calcium chloride is added to form insoluble calcium alginate; (5) The solid calcium alginate is washed with an acid solution; (6) The acidified calcium alginate is neutralized with sodium carbonate to form sodium alginate; and (7) The sodium alginate is filtered and dried.

In the nachos dish, the sodium alginate is found in the guacamole, probably to add body and consistency. There are no safety issues for sodium alginate.

Trehalose (Tremalose, Mycose):

Before working on this book, I had never heard of this additive. Not a big surprise since it did not enter the processed food marketplace until after 2001. The compound was first isolated in 1832 from rye fungus and then again in 1859 where it was found in the cocoons of the Larinus maculates beetle in the desert areas of Turkey. In the cocoons was a substance called trehala manna. "Trehala" is a Turkish word for the sweet substance inside the cocoon. The "-ose" suffix tells you that this substance is a type of sugar (think suc<u>rose</u> and fruct<u>ose</u>). It's an unusual derivative of glucose. The word "manna" is interesting in that some people think that the manna from heaven mentioned in the Old Testament actually refers to the edible substance found in the cocoons of the desert beetles of the Middle East.

This is such an interesting substance that I'm going to spend a little bit of time talking about it. Although trehalose has been known for a long time, it was too expensive to be used as a food additive or industrial chemical. Prior to the 2000s, trehalose cost $7000 per kilogram (2.2 lbs) or $3182/lb. Japanese scientists figured out a way to make it efficiently and cheaply from starch using a method called enzymatic conversion. Today it costs only $3 per kilogram to produce, a 1061 times reduction in price, and you can buy it on Amazon.

Trehalose has also been found in bacteria, mushrooms (10 to 25% by weight), and foods with brewer's yeast, as well as in animals such as shrimp, grasshoppers, locusts, butterflies, and bees where it functions as a blood sugar. When there is a drought, trehalose forms a gel phase in cells undergoing dehydration, thereby protecting the cells from drought and high salt conditions. It also protects against protein degradation during severe environmental stresses. As the blood sugar in insects, it is estimated that they can use it twice as efficiently as the glucose in mammals. Additionally, trehalose is the secret ingredient in the "resurrection plant", which can survive months without water. With trehalose in their cells, some plants can lose 95% of their water content and still survive!

Oddly enough, our bodies have an enzyme, trehalase, that can break down trehalose into glucose. Some people think that this is a throwback to a pre-historical time when humans may have eaten insects as a staple food.

Trehalose is sweet, but it's only 45% as sweet at table sugar, and, strangely, the sweetness decreases with increasing concentration in water. On the plus side, it doesn't raise blood sugar levels as high as glucose. Just as in insects, this sugar can lower the freezing point of water, so it's sometimes used in frozen desserts. Its other properties are (1) functions as a texturizer, (2) the ability to mask bitterness and food odors, (3) enhancing saltiness and highlighting fruity flavors, and (4) preserving food. The food industry sometimes lists it as a natural flavor.

Trehalose is on the FDA's GRAS list, so it's considered safe to consume. If trehalose sounds too good to be true, you're right. There's a scary side to it. Some studies have indicated that this chemical can boost the production of Clostridium difficile bacteria by more than 500-fold. This bacterium is the most common cause of health-care infections in hospitals. It may be a coincidence, but deaths due to C. diff increased after the introduction of trehalose in the marketplace in 2000.

Finally, where does trehalose show up in the nachos dish? It's in the refried beans. Why a more common and cheaper sugar, like sucrose or dextrose, is used in the bean preparation, I have no idea!

Taco Bell's Wild Strawberry Freeze™ (190 Calories for 20oz)

Ingredients:

Wild Strawberry Freeze™ (16 oz): Wild Strawberry Freeze Base: High Fructose Corn Syrup(1), Water(2), Citric Acid(3), Natural Flavor(4), Yucca Mohave Extract(5), Sodium Benzoate(6), Red 40(7), Gum Arabic(8), Glycerol Ester of Rosin(9), Mixed Triglycerides(10).

Candy Seeds: Dextrose(11), Corn Syrup(12), Calcium Stearate(13), Tapioca Dextrin(14), Confectioner's Glaze(15), Carnauba Wax(16), Artificial Flavors(17), Citric Acid(18), Artificial Colors including FD&C Red #40(19) and Blue 1(20).

Ingredient Assignments and the Degree of Industrialization (Processing)

Artificial Flavors (synthetic) (17) Extremely; **Blue 1** (synthetic) (20) Extremely; **Calcium Stearate** (synthetic) (13) Extremely; **Carnauba Wax** (16) Moderately; **Citric Acid** (synthetic) (3,18) Extremely; **Confectioner's Glaze** (15) Moderately; **Corn Syrup** (12) Highly; **Dextrose** (synthetic) (11) Extremely; **Glycerol Ester of Rosin** (synthetic) (9) Extremely; **High Fructose Corn Syrup** (synthetic) (1) Extremely; **Mixed Triglycerides** (10) Highly; **Natural Flavors**

(4) Moderately; **Red 40** (synthetic) (7,19) Extremely; **Sodium Benzoate** (synthetic) (6) Extremely; **Tapioca Dextrin** (14) Highly.

Reporting the Processed Food Index (PFI) for This Food Item

# OF UNPROCESSED INGREDIENTS:	1
# OF LIGHTLY PROCESSED INGREDIENTS:	4
# OF MODERATELY PROCESSED INGREDIENTS:	3
# OF HIGHLY PROCESSED INGREDIENTS:	3
# OF EXTREMELY PROCESSED INGREDIENTS:	9
TOTAL # OF INGREDIENTS:	20
# OF UNIQUE INGREDIENTS:	20 (100%)
PROCESSED FOOD INDEX:	55 (Highly Industrialized)

Commentary

Of course, this menu item being a beverage, we don't expect to see a high number of ingredients, as in the other Taco Bell offerings. There are only 20 ingredients, but notice that 100% of them are unique. Of the 20 ingredients, 15 of them or 75% are significantly processed. That's why the PFI score of 55 is higher than usual for a beverage. This drink is considerably far from being natural. I even challenge you to find anything in the ingredient list that you would consider natural, i.e., pure and straight from nature. Oddly, in this "wild strawberry" drink, strawberries are not one of the ingredients! That's where the "natural" flavor comes in ... maybe the flavor was chemically extracted from strawberries. Sadly, we don't know whether strawberries were involved or not since the food company does not have to reveal what's in the "natural" flavor.

Here is part of the description of this product from the Taco Bell website. They put an incredibly weird, marketing spin on this mostly synthetic beverage.

"The Wild Strawberry Freeze is a go-getter. It doesn't rest until the job is done. An entrepreneur. A hustler. In this context, 'wild' is most closely synonymous with 'crazy'. Why? People call it 'wild' because they can't understand how it can be so motivated. It has to be crazy, right? Nope. Wild Strawberry Freeze just lives life to fiercest. Wait, are we still talking about a freeze? Absolutely. It's frozen. It tastes like real wild strawberries. However, what's a freeze without a little personality? Just a Strawberry Freeze. The wild makes it worth your while. Find out if you can face the fierce flavor for yourself at participating locations ... Freezes are made with artificial flavors and contain no fruit juice."

After all the marketing hype, Taco Bell is strikingly honest at the end of the description where they admit that the beverage is made with artificial flavors and has no fruit juice in it. Now that's "wild!"

We find some unusual ingredients listed in the Candy Seeds section. To make this beverage stand out from the freezes offered by other restaurants, Taco Bell incorporates black-colored, round, candy pieces into it. There are some unusual ingredients in those candies and those are discussed in detail below.

Lastly, the calorie content at 190 is not too bad. That's less than the 240 calories in a 20-oz Coca Cola.

The highlighted ingredients in Taco Bell's Wild Strawberry Freeze™ are:

- Carnauba Wax

- Confectioner's Glaze

- Glycerol Ester of Rosin

Carnauba Wax (Brazil Wax, Palm Wax):

It may sound rather strange that wax is a food ingredient, but it does have a variety of applications in the food industry. Carnauba wax only comes from one place: the northeastern states of Brazil. It forms on the leaves of the carnauba palm during hot and dry weather where it helps to retain moisture. Farm workers beat the leaves to flake it off and then it's collected for later treatment. The raw wax is stored as hard, yellow-brown flakes. At a later stage, the wax is refined into different grades and may be bleached to reduce the color. It is sold as flakes or a powder.

Carnauba wax has a very high melting point, so it's very stable. It's insoluble in water and alcohol, so it holds up in food products. It is considered non-toxic, although some people may be allergic or hypersensitive to it. In the food industry, it's applied to candy and gum tablets to provide a shiny, protective coating, e.g., in M&Ms. It's also found in fruit snacks and gummy candies where it provides texture and stability. In the Wild Strawberry Freeze™, it's the coating on the "candy seeds." When ingested, carnauba wax just passes through the body since it can't be broken down. The wax has a number of uses in other industries such as car polish, shoe polish, and finishes on musical instruments, where it creates a glossy appearance.

Earlier I challenged you to find a "natural" ingredient in this menu item. Actually, the carnauba wax and the following additive come closest to that description.

Confectioner's Glaze (Resinous Glaze, Pure Food Glaze, Food-grade Shellac):

The word "confectioner" refers to a candy maker, and the word glaze refers to a glossy coating. So, confectioner's glaze is a material that provides the "candy seeds" in the beverage with a glossy, protective coating. Where does this material come from? You may be a bit surprised. It's derived from the natural material called sticklac obtained in southeast Asia. That's a resin excreted by the lac beetle when it's building a hard, water-proof cocoon in the trees that it lives in. Workers scrape the resin from the branches of the trees. Unfortunately, during the harvesting process, a large number of the insects are killed causing about 25% of the sticklac to be composed of insect body parts. Obviously, any foods that contain confectioner's glaze are not suitable for vegetarians or vegans.

The sticklac is dissolved in alcohol and used as food-grade shellac. Examples of candies that use the shellac are candy corn, Hershey's Whoppers and Milk Duds, Nestlé's Goobers, Junior Mints, and jelly beans.

There are no safety issues associated with this additive.

Glycerol Ester of Rosin (Glyceryl Abietate, Ester Gum):

This substance is a rather strange sounding additive, but it comes from a natural source. Rosin, also known as resin, is a translucent yellowish to dark brown material obtained from the stumps or sap of various pine trees. The mixture, known as glycerol ester of rosin, is prepared by reacting refined wood rosin, obtained by solvent extraction, with food-grade glycerin. The raw product is purified by steam stripping. The final material is a yellow to pale amber solid. It's composed chiefly of abietic acid and related compounds. Its purpose is to keep oils in suspension in water, particularly in beverages such as soda, lemonades, vitamin-enhanced water, and sports drinks. It's also used in chewing gums as a base. It can also extend shelf-life for foods and beverages. The FDA approved the use of this additive in the early 1960s.

In the case of the Wild Strawberry Freeze™, the glycerol ester of rosin probably serves to suspend the natural flavors, probably strawberry-flavored oils, so they don't separate out of the water mixture. Otherwise, the flavoring oils would just simply float on top of the beverage. That's why the additive is described as a stabilizer.

This additive is only present in foods in very small amounts. The upper limit is 100 parts of rosin per million parts or about 60 mg in a 20-oz beverage. Although synthetically made, there are currently no safety concerns associated with this additive.

What About a Whole Taco Bell Meal?

All of the food and beverage evaluations so far have dealt with individual menu items. Let's take a quick look at a possible Taco Bell meal: Beef Burrito, Grilled Chicken Grande Nachos, and Wild Strawberry Freeze™.

BEEF BURRITO:

Flour Tortilla: Enriched Wheat Flour(1), Water(2), Vegetable Shortening (Soybean(3), Hydrogenated Soybean(4) and/or Cottonseed Oil(5)), Sugar(6), Salt(7), Leavening (Baking Soda(8), Sodium Acid Pyrophosphate(9)), Molasses(10), Dough Conditioner (Fumaric Acid(11), Distilled Monoglycerides(12), Enzymes(13), Vital Wheat Gluten(14), Cellulose Gum(15), Wheat Starch(16), Calcium Carbonate(17)), Calcium Propionate(18), Sorbic Acid(19), and/or Potassium Sorbate(20).

Seasoned Rice: Enriched Long Grain Rice(21), Water(22), Canola Oil(23), Seasoning (Maltodextrin(24), Salt(25), Natural Flavors(26), Tomato Powder(27), Sugar(28), Garlic Powder(29), Spices(30), Onion(31), Tomato(32), Red(33) and Green Bell Peppers(34), Citric Acid(35), Paprika(36), Onion Powder(37), Paprika(38), Disodium Guanylate(39) and Inosinate(40), Torula Yeast(41).

Nacho Cheese Sauce: Nonfat Milk(42), Cheese Whey(43), Water(44), Vegetable Oil (Canola Oil(45), Soybean Oil(46)), Modified Food Starch(47), Maltodextrin(48), Natural Flavors(49), Salt(50), Dipotassium Phosphate(51), Jalapeno Puree(52), Vinegar(53), Lactic Acid(54), Cellulose Gum(55), Potassium Citrate(56), Sodium Stearoyl Lactylate(57), Citric Acid(58), Annatto(59) and Oleoresin Paprika(60).

Creamy Jalapeno Sauce: Soybean Oil(61), Water(62), Vinegar(63), Jalapeno Peppers(64), Buttermilk(65), Spices(66), Cage-Free Egg Yolk(67), 2% or less of: Dextrose(68), Chili Powder(69), Salt(70), Natural Flavors(71), Sugar(72), Onion Powder(73), Paprika(74),

Minced Onion(75), Garlic Powder(76), Cocoa Powder(77) Processed With Alkali, Dried Onion(78), Glucono-Delta-Lactone(79), Modified Food Starch(80), Xanthan Gum(81), Potassium Sorbate(82), Sodium Benzoate(83), Sorbic Acid(84), Disodium Inosinate(85) and Disodium Guanylate(86), Lactic Acid(87), Propylene Glycol Alginate(88), Citric Acid(89), Acetic Acid(90), Calcium Disodium EDTA(91).

Seasoned Beef: Beef(92), Water(93), Seasoning [Cellulose(94), Chili Pepper(95), Maltodextrin(96), Salt(97), Oats(98), Soy Lecithin(99), Spices(100), Tomato Powder(101), Sugar(102), Onion Powder(103), Citric Acid(104), Natural Flavors(105) (including Smoke Flavor(106)), Torula Yeast(107), Cocoa(108), Disodium Inosinate(109) & Guanylate(110), Dextrose(111), Lactic Acid(112), Modified Corn Starch(113)], Salt(114), Sodium Phosphates(115).

GRILLED CHICKEN GRANDE NACHOS:

Nacho Cheese Sauce: Nonfat Milk(116), Cheese Whey(117), Water(118), Vegetable Oil (Canola Oil(119), Soybean Oil(120)), Modified Food Starch(121), Maltodextrin(122), Natural Flavors(123), Salt(124), Dipotassium Phosphate(125), Jalapeno Puree(126), Vinegar(127), Lactic Acid(128), Cellulose Gum(129), Potassium Citrate(130), Sodium Stearoyl Lactylate(131), Citric Acid(132), Annatto(133) and Oleoresin Paprika(134).

Nacho Chips: White Ground Corn(135), Water(136), 2% Or Less Of: Calcium Propionate(137), Fumaric Acid(138), Sorbic Acid(139), Cellulose Gum(140), Sodium Propionate(141), Salt(142). Prepared in Canola(143) and/or Vegetable Oil(144).

Refried Beans: Pinto Beans(145), Soybean Oil(146), Seasoning (Salt(147), Sugar(148), Spice(149), Beet Powder(150), Natural Flavors(151), Sunflower Oil(152), Maltodextrin(153), Corn Flour(154), Trehalose(155), Modified Corn Starch(156)).

Guacamole: Avocado(157), Water(158), Tomato(159), Onion(160), Jalapeno(161), Salt(162), Cilantro(163), Lemon(164) Or Lime Juice(165), Ascorbic(166) or Erythorbic Acid(167), Xanthan Gum(168), Sodium Alginate(169).

Reduced-Fat Sour Cream: Milk(170), Cream(171), Modified Corn Starch(172), Lactic Acid(173), Maltodextrin(174), Citric Acid(175), Sodium Phosphate(176), Natural Flavor(177), Cellulose Gel(178), Potassium Sorbate(179), Cellulose Gum(180), Guar Gum(181), Locust Bean Gum(182), Carrageenan(183), Vitamin A(184).

Three Cheese Blend: Low-Moisture Part-Skim Mozzarella Cheese, Cheddar Cheese, Pasteurized Process Monterey Jack And American Cheese With Peppers (Cultured Milk(185), Cultured Part-Skim Milk(186), Water(187), Cream(188), Salt(189), Sodium Citrate(190), Jalapeno Peppers(191), Sodium Phosphate(192), Lactic Acid(193), Sorbic Acid(194), Color Added: Annatto(195) and Paprika Extract Blend(196), Enzymes(197), Anticaking Agents (Potato Starch(198), Cornstarch(199), Powdered Cellulose(200))).

Tomatoes: Fresh Tomatoes(201).

Grilled Chicken: White Meat Chicken with Rib Meat(202), Water(203), Seasoning (Modified Potato Starch(204), Salt(205), Yeast Extract(206), Spices(207), Sugar(208), Citric Acid(209), Disodium Inosinate(210) and Guanylate(211), Torula Yeast(212), Maltodextrin(213), Dextrose(214), Natural Flavors(215), and 2% or less of Soybean Oil(216) added as a processing aid), Sodium Phosphates(217).

WILD STRAWBERRY FREEZE™:

Wild Strawberry Freeze™ (20 oz): Wild Strawberry Freeze Base: High Fructose Corn Syrup(218), Water(219), Citric Acid(220), Natural Flavor(221), Yucca Mohave Extract(222), Sodium Benzoate(223), Red 40(224), Gum Arabic(225), Glycerol Ester of Rosin(226), Mixed Triglycerides(227).

Candy Seeds: Dextrose(228), Corn Syrup(229), Calcium Stearate(230), Tapioca Dextrin(231), Confectioner's Glaze(232), Carnauba Wax(233), Artificial Flavors(234), Citric Acid(235), Artificial Colors including FD&C Red #40(236) and Blue 1(237).

So, here we have a 3-component meal with a staggering 237 ingredients. The ingredients highlighted in gray have been deemed significantly processed. There are 152 of them in this meal. That's 64.1% of the total or, another way of putting it, about 6 out of 10 of the ingredients in this meal are significantly processed. Or, you could say that more than 50% of this meal deviates from natural sources of food. Another way to look at this meal is to observe that about 94 of the 237 ingredients, or 40%, are synthetic or man-made. A large portion of the ingredients for this great-tasting meal weren't even available 150 years ago. The modern food system created the ingredients and the applications to use them.

How many calories are we looking at for this meal? The beef burrito brings 460 calories. The Grilled Chicken Grande Nachos contributes 1070 calories. Finally, the Wild Strawberry Freeze™ provides 190 calories. The sum total for the meal is 1720 calories. If 2000 calories per day is the target for the average adult American, the caloric content of this meal amounts to 86% of the recommended daily amount. That leaves a meager number of calories for the rest of the day. Just imagine the total caloric intake for the day if someone chose to eat two additional high-calorie, fast-food meals?

CHAPTER 14

KEY TAKEAWAYS IN PART II

1. A menu item with a high number of ingredients is the first clue that you've encountered a very processed food.

2. It's important to understand that every additive that winds up in a commercial food is there for a purpose because it provides a specific functionality to the food.

3. Highly industrialized foods are characterized by (1) a high percentage of unique ingredients; (2) a high percentage of highly processed ingredients; (3) a high processed food index (PFI), typically over 40; and (4) a large number of synthetic ingredients.

4. Keep in mind that behind every commercial food there is a ton of research and testing carried out prior to market launch to assure an appealing, safe, commercial product at the lowest possible cost for the company.

5. The sodium content on a food label will not only reflect the presence of salt (sodium chloride) in the food, but also any other ingredient that contains sodium.

6. Food manufacturers want their products to last as long as possible on store shelves … that's called shelf life.

7. There are basically two kinds of preservatives: (1) antioxidants and (2) antimicrobials. The former remove oxygen which can causes fats and oils to go rancid. The latter inhibits the formation of bacteria, mold, and yeast.

8. The FDA allows the use of a handful of acronyms to represent ingredients. Unfortunately, acronyms are not very

useful to consumers, who likely won't know what they stand for.

9. There are hundreds, if not thousands, of natural flavor ingredients. Companies protect them as trade secrets (proprietary formulas), and the FDA allows them to hide their identities on food labels.

10. It's interesting to note that regulatory agencies around the world do not agree with one another regarding the safety of food dyes. For example, in Europe, certain dyes (synthetic food colorings) can be used in foods, but the products are required to carry a warning label which could say "may have an adverse effect on activity and attention in children." There are no such label requirements in the USA.

11. Ultra-processed foods are engineered by food scientists. They require a host of synthetic chemicals to add flavor, color, texture, protect from spoilage, protect from unfavorable changes over time, hold incompatible components together like oil and water, alter and fix acidity levels, and replace nutrients that were removed from the natural food during processing.

PART III

HEALTH STUDIES & SCORING SYSTEMS

CHAPTER 15

HEALTH STUDIES OF ULTRA-PROCESSED FOODS

"... we are now seeing in real time the negative global impact of this kind of diet [Standard American Diet]. Every country that has changed its dietary traditions to incorporate those in fashion in North America also sees a rapid increase in its rates of obesity, colon and prostate cancers, and heart disease – all diseases that for them were until recently relatively rare."

— *"Foods to Fight Cancer" a book by Richard Béliveau, Ph.D. and Dr. Denis Gingras.*

Earlier, in the Introduction to this book, I mentioned that I wasn't going to get into the nutritional and health aspects of individual fast foods. That's an important topic, but the scope is so large that I would be writing a book within a book. However, I do want to discuss the health aspects of diets high in ultra-processed foods (UPF). A Google search on any given day would reveal that this topic is being constantly addressed in the media. In this chapter, I want to delve into the topic by looking at the scientific literature to see what it has to say on the subject.

Most people, with only a nominal understanding of health and nutrition, could agree that consuming a diet rich in ultra-processed foods might be devastating to their overall well- being. A diet heavy in ultra-processed foods exposes people to high consumptions of sugar, oil, salt, and myriad industrial chemicals potentially leading to disabling illnesses and disorders like heart disease, type 2 diabetes, kidney disease, metabolic syndrome, Alzheimer's, and cancer, to name just a few. Of course, the phenomenal rise of obesity in the USA has greatly contributed to the development of many of these chronic medical conditions. Certainly, these health issues have existed since the beginning of humankind, but the risk factors have gone up considerably in the last century, likely due to two factors: mortality and diet. The tables below, based on data obtained from the Centers for Disease Control (CDC) for the last 50 years, show mortality (death) statistics. **TABLE A** shows the death rate from 1970 to 2019 and **TABLE B** shows the death statistics for four major diseases in the USA: heart disease, cancer, stroke, and diabetes.

Further Readings:
Leading Causes of Death, 1900 - 1998

https://www.cdc.gov/nchs/data/dvs/lead1900_98.pdf

Deaths: Leading Causes for 2000

https://www.cdc.gov/nchs/data/nvsr/nvsr50/nvsr50_16.pdf

Deaths: Final Data for 2010

https://tinyurl.com/25343xov

Deaths: Leading Causes for 2019

https://www.cdc.gov/nchs/data/nvsr/nvsr70/nvsr70-09-tables-508.pdf

TABLE A: DEATH RATE FROM 1970 TO 2019 IN USA

YEAR	POPULATION (MILLIONS)	TOTAL DEATHS	DEATH RATE (%)
1970	210	1921031	0.91
1980	229	1989841	0.87
1990	252	2148463	0.85
2000	282	2403351	0.85
2010	309	2468435	0.80
2019	331	2854838	0.86

Source: Author Generated Table

TABLE B: % DEATHS FROM SOME CHRONIC DISEASES IN USA, 1970 – 2019

Year	Heart Disease	% Total Deaths	Cancer	% Total Deaths	Stroke	% Total Deaths	Diabetes	% Total Deaths
1970	735542	38.29	330730	17.22	207166	10.78	38324	1.99
1980	761085	38.25	416509	20.93	170225	8.55	34851	1.75
1990	720058	33.52	505322	23.52	144088	6.71	47664	2.22
2000	710760	29.57	553091	23.01	167661	6.98	69301	2.88
2010	597689	24.21	574743	23.28	129476	5.25	69071	2.80
2019	659041	23.09	599601	21.00	150005	5.25	87647	3.07

Source: Author Generated Table

Some interesting observations can be made from the data above. Notice in **Table A** that the death rate in the USA has been fairly consistent for the last 50 years. The high was 0.91% in 1970, probably due to the last years of the Vietnam War. The low (0.80%) was in 2010.

How have the % of deaths due to certain diseases changed over the last 50 years in the USA? The deaths due to heart disease have declined significantly due to early detection and improved medical treatment. On the other hand, deaths due to cancer have steadily increased. For strokes, in parallel with heart disease, the deaths have dropped by about 50%, again due to improved medical treatment (particularly emergency care). For diabetes (type 2), a metabolic disease strongly associated with diet and lifestyle, has dramatically risen along with the death percentage.

Let's look at the two major factors affecting these statistics. First, there's mortality, which is a measure of the frequency of death in a population. As much as we might not want to think about it, we all have to die of something eventually. Some diseases, like heart disease, usually take a long time to develop, so, obviously, the longer we live the higher the risk of dying from that disease. Hence, age is a major determinant as regards the onset of a disease and its fatal effects. In the last century, how has the USA fared in terms of average age of death (life span)? Take a look at TABLE C which is based on data from the United Nations.

TABLE C: USA LONGEVITY IN YEARS FROM 1970 TO 2020

YEAR	LONGEVITY IN YEARS
1970	70.8
1980	73.7
1990	75.2
2000	76.8
2010	78.5
2020	78.9

Source: Author Generated Table

Further Readings:
Chart of US Life Expectancy from 1950 to 2022

https://www.macrotrends.net/countries/USA/united-states/life-expectancy

Notice the jump in longevity of 8.1 years from 1970 to 2020. That number would be a little higher if it weren't for the onset of the COVID-19 pandemic in 2020. So, we Americans, on average, are steadily living longer. Why is that happening despite dietary threats? There are numerous explanations, but the two primary ones are healthcare and nutrition. Since the late 19th century, advances in modern medicine have drastically reduced the number of people dying from specific causes -- reduction of infectious diseases, control of parasites, improved sanitation, decreases in pre-natal and post-natal deaths, lifesaving medicines and medical procedures, improved monitoring devices and testing, etc. [Note that in the year 1900 about 3 times more people died of infectious diseases than heart disease, the #1 killer today.] Also, longevity and healthcare has dramatically improved as a consequence of the ever-increasing knowledge about nutrition. The existence of vitamins was virtually unknown in the 19th century. There was a rudimentary understanding of the role of minerals. And, of course, the discovery of phytochemicals, anti-oxidants, the microbiome (microbial environment of the gut), and the essential properties of fiber for gut health were not known. We now have a great deal of knowledge on how to sustain health, not only in terms of beneficial foods, but also in terms of healthy lifestyles, including adequate exercise and stress control. In effect, the last one-and-a-half centuries of medical advancements have taken us from the lethal threats of acute illnesses and accidents to the slow, progressive takedowns by chronic diseases. Of course, the threats of infectious diseases, like

COVID-19 from 2020 onwards, are changing the risks and statistics of the future.

Thanks to advances in science, technology, and modern medicine, the average American, man or woman, is living longer. Isn't that wonderful? Well, maybe. The average life span in the USA is a quantitative, statistical value. What it fails to take into account is the quality of life ... how do your final years pan out? Quantitative vs. qualitative: that's an interesting choice. Can you have both? Sure, but not without making some important personal decisions and conscious efforts. You can have <u>both</u> a long life and a reasonably healthy one (lest we forget, we will all wind up dying of something). Strangely, when people are asked about this topic, more people are singularly focused on lifespan, rather than quality of life; they are unwilling to give up their personal pleasures and comforts in exchange for additional, healthy years of living. A typical answer is "Why do I want to live longer if I have to give up or reduce the things I enjoy? It's much better to live a shorter, enjoyable life to the max than to live a longer, less satisfying life." Of course, what this opinion fails to take into account is the pain, suffering, and misery that often accompanies the final days or years of an unhealthy life ... these can be days or years of agony unless you're lucky enough to be taken quietly in the night or are killed in an accident. I would be interested in reading surveys which queried people on their deathbed to see whether they had changed their minds about quantity vs. quality of life.

Interestingly, this question about personal life expectancy vs. end-of-life quality is an on-going one in the fields of health, psychology, sociology, and economics. What does it cost to garner a person a few more years of life? Is the cost worth it just to extend life without a return of quality (healthy) years? Is it worth it for you to live a few more years where you are bed-ridden in a nursing home with very little or no ability to care for yourself? A couple of theoretical models try to answer these questions. The Quality-

Adjusted Life Year (QALY) is one model. It takes into account the influence of health issues on the quality and quantity of life. A single QALY equals one year of very good health. Less than good health is represented by a QALY score of under one. Of course, if your QALY score is zero, you're likely dead or in a coma. This scoring system has been used to evaluate the benefits of health interventions (surgeries, drugs, etc.). If a very expensive operation is expected to yield only a 0.1 QALY, is it worth doing it? On the other hand, if the same operation affords another person a QALY score of 4, maybe they should grab the opportunity, since they could live another 4 years in relatively decent health. The really difficult aspect of this model is assessing the improvement associated with the operation, drug, or lifestyle change.

Another model is called the Healthy Life Years (HLY) indicator. It's also called the disability-free life expectancy index. This model, originating in Europe, measures both mortality (death) and morbidity (disease). The model applies only to populations, rather than individuals. Assigned to a specific age, the HLY provides a quantitative number of the remaining years of life without the burden of a disability. The HLY is based upon population life tables and self-perceived disability using personal surveys.

Details of either model are beyond the scope of this book. The models are data driven, mathematically based, and can be quite complex. The reader is encouraged to research QALY and HLY if more information is desired.

As regards the quality of life, the focus of this book is on ultra-processed foods, particularly in the fast-food industry. The rest of this chapter explores recent, scientific studies that reveal a relationship between the elevated consumption of ultra-processed foods and resulting poor health outcomes, which ultimately affect both the quantity and quality of life.

The Early Research

In the last few years, as I scanned food-related news headlines and articles from the scientific literature, the name of one scientist kept appearing over and over again as regards the relationship between nutrition and highly processed foods: Carlos Monteiro. He is the world's foremost researcher in this area. Dr. Monteiro is a professor of nutrition and public health at the University of Sao Paulo in Brazil and chairs the Center for Epidemiological Research in Nutrition and Health. He is credited with coining the now common expression "ultra-processed food."

It only makes sense to start with his description of ultra-processed foods, which was taken from an interview in March 2019. His words are slightly modified for syntax and punctuation. "They are a formulation of industrial ingredients -- very low-cost ingredients. And this was possible by the development of food technology in recent years. So, ultra-processed foods actually result from the fractioning of whole foods into components and includes the recombination of these components ... in a way, ultra-processed foods result from advances in food science and food technology. But it is not an advance in terms of human health because our bodies simply aren't prepared to be fed with these formulations; these formulations lack the food matrix. ... So, essentially ultra-processed foods are inventions of modern food technology that allow huge profits, but are harmful to our bodies."

Let's break down Dr. Monteiro's statement using the example of a whole food, corn. Kernels of corn can be fractionated into component parts, one of which is starch. The starch can then be broken down into the sugar, glucose. In turn, the glucose can be chemically converted into the sugar, fructose. Specific combinations of glucose and fructose produce various corn syrups, the most famous of which is high-fructose corn syrup (HFCS), ubiquitous in sweetened beverages (note that HFCS is not found in nature). So, starting with a whole food, corn, a bunch of food ingredients can be prepared by breaking it apart, then recombining (or modifying) the breakdown products. According to Dr. Monteiro, these new food ingredients, inventions of modern food technology, and their combinations are not real foods. As a consequence, they wind up adversely affecting our health in a number of ways.

Dr. Monteiro goes on to say, "We know for instance, that when you do these formulations --combining sugar, salt, fats, and additives that are particular flavors, you produce products that tend to be consumed in excess. So, these products and the formulation of these products actually aim to fool our bodies in terms of making our bodies consume more than we need. So, in a way, we lose the ability to control the amount of food we need because we are not really consuming food. We are consuming hyper-palatable formulations."

Here, Dr. Monteiro brings up the concept of hyper-palatable foods. Ultra-processed foods, in combination with sugar, salt, and fat, plus industrially produced flavors, create foods that we can't resist. So, we wind up over-consuming those types of foods, which leads to unfavorable health outcomes.

Again, in the words of Dr. Monteiro, "I mentioned before that that they [ultra-processed foods] are full of additives, and, when we say full, we are saying that dozens and dozens of new additives every year enter into the food supply. And, in some cases, like emulsifiers, for instance, [they are] very common in ultra-processed foods. We know that emulsifiers can affect the impermeability of our intestinal cells. So, then, we again lose the ability to control what goes in our bodies because these emulsifiers destroy some protection we have against the absorption of some molecules. Artificial sweeteners [are] another common additive used in ultra-processed foods; we know today that they have a big effect on our bodies. So, in a way, ultra-processed food represents a problem because they contain intrinsically imbalanced macronutrients -- too much sugar, too much unhealthy fats, too much sodium. But, at the same time, they include things that are completely strange to our body like the additives. So, then this explains why we see in epidemiological studies that the more a person consumes ultra-processed foods, the higher the risk of several chronic diseases including obesity, cardiovascular diseases, [and] certain types of cancer … they represent a big public health problem today."

In this last statement, Dr. Monteiro points out that another aspect of ultra-processed foods is the use of additives. Although, food additives provide functional value to the food (preservation, texture, uniformity, taste, etc.), many of these foreign substances adversely affect the healthy functioning of the body. Ultimately, people who consume large quantities of ultra-processed foods assume a higher risk of developing chronic diseases. And that's the subject of this chapter.

Dr. Monteiro addresses a question regarding the benefits of technology as applied to food. "The question of ultra-processed food is that they are really no longer foods, in a way. They don't preserve the food matrix, but at same time they taste and they look like real food. And I think that that's the origin of all the problems we have. So, perhaps we need to admit that we need to be much, much more careful when we apply technology to the food in the food system."

Here Dr. Monteiro is questioning the advantages of altering food using technological know-how. Technical innovation in the computer industry works great because efficiency increases and cost decreases; that's fantastic for the consumer, but computer products don't directly affect our health. Radical modifications of whole foods convert them into substances that our bodies are unprepared to handle. Consequently, the overconsumption of industrial chemicals puts our bodies at risk.

Further Readings:
The Leading Voices in Food Podcast Series, Episode 24: Carlos Monteiro on the Dangers of Ultra-processed Foods, Duke Sanford World Food Policy Center

https://tinyurl.com/28amnu27

Next, let's take a look at some of the research studies that Dr. Monteiro and his colleagues have conducted and reported on.

The Origin of the Term 'Ultra-Processed Food' in an Early Monteiro Paper

The impact of highly processed foods on public health was first brought to light in 2009 by Dr. Carlos Monteiro. In an article published in Public Health Nutrition, he pointed out that public health education focused on the benefits of specific nutrients, like vitamins, and the detriments of some food components, such as saturated fat. Furthermore, national food guidelines promoted the consumption of certain foods over others. However, he pointed out that public health advocates failed to consider the effects of food processing in their recommendations, even though highly processed foods (i.e., many pre-prepared foods) were linked to rising rates of obesity and chronic diseases.

Dr. Monteiro was not criticizing food modification or processing in general, which can improve the availability and quality of food, but he was addressing the type and intensity of processed foods. For clarification, he proposed dividing processed foods/drinks into three groups. I won't describe the proposed groups in detail here because later he expanded the system to four groups, and these I'll discuss in the next section. I'll just say that the foods in the third group were classified as ultra-processed foods. He described the latter as "confections of group 2 ingredients, typically combined with sophisticated use of additives, to make them edible, palatable, and habit-forming ... designed to be ready-to-eat or ready-to-heat, ... and often consumed alone or in combination ... Ultra-processed products are typically branded, distributed internationally and globally, heavily advertised and marketed, and very profitable." He also pointed out that "traditional diets wholly or mainly made up from unprocessed and minimally processed foods ... usually have adequate nutrient and energy density when they contain varied combination(s) of plant foods (grains, vegetables, pulses, fruits, nuts), only moderate quantities of animal foods, and little salt." He emphatically stated, "Diets that include a lot of ultra-processed foods are intrinsically nutritionally unbalanced and intrinsically harmful to health." He criticized food manufacturers for labeling some products as 'premium' suggesting that by reducing some negative components, like fat, trans fat, sugar, and salt and boosting positive components like micronutrients, vegetables, fruits, and nuts would magically transform their products into healthy foods.

In the next section, I describe Dr. Monteiro's food classification system still in use today.

Further Readings:
"Nutrition and health. The Issue is Not Food, Nor Nutrients, So Much as Processing"

Monteiro, et. al., 2009.

http://www.wphna.org/htdocs/downloadsdec2012/2009_PHN_Monteiro.pdf

The NOVA System

A paper by Dr. Monteiro was published in the World Nutrition Journal in 2016 entitled "NOVA. The Star Shines Bright." When I first saw the term NOVA, I thought it was an acronym, but it's not.

It's a fully capitalized form of the word "nova," a star that rapidly increases its luminescence. Apparently, Dr. Monteiro's team wanted to develop a food classification system that would shed light on commercial foods.

Soon after its introduction, the new classification was rapidly adopted by other health and nutrition researchers, referred to in many other scientific papers, and is frequently referenced by health organizations around the world.

So, what is NOVA? As stated in the article, "NOVA is the food classification that categorizes foods according to the extent and purpose of food processing, rather than in terms of nutrients." The system divides all commercial foods into four distinct groups. Each group contains specific foods representative of the different types of processing used. Also, from the article, "Food processing as identified by NOVA involves physical, biological, and chemical processes that occur after foods are separated from nature and before they are consumed or used in the preparation of dishes and meals."

Group 1 Foods

This group includes any unprocessed plant foods (seeds, fruits, leaves, stems, roots, bark), animal foods (muscle, offal, excreted products like milk and eggs), or fungi and algae. These foods are typically considered natural and only processed to remove inedible or unwanted parts using a variety of processing methods such as drying, crushing, grinding, filtering, roasting, boiling, pasteurizing, cooling, non-alcoholic fermentation, etc. causing negligible changes to the foods. The primary purpose of these processes is to extend the food's life, but no additional food substances are added. Some examples of Group 1 foods include fruits, leafy greens, grains, legumes, lentils, starchy roots and tubers, mushrooms, meat, poultry, fish, seafood, eggs, milk, flours or grits (made from grains), herbs and spices, plain yogurt, natural beverages (tea, coffee) and water. Also, combinations of two or more of these food items are also members of Group 1, as well as fortified foods (containing added vitamins/minerals).

Group 2 Foods

This group is composed of processed culinary ingredients which are used in home and restaurant kitchens. They are derived from the Group 1 foods or from nature by various techniques of processing, e.g., pressing, refining, grinding, milling, and spray drying. These materials are rarely used as standalone ingredients but instead enhance the foods of Group 1. Examples of foods in this group include salt, extracted natural sugar and molasses, honey, maple syrup, vegetable oils obtained directly from plants by pressing, butter, lard, and starches extracted from corn and other plants. Two or more items from this group may be combined together as in iodized salt, salted butter, and acetic acid from the fermentation of wine. Some Group 2 items may contain preservatives to maintain the product's original properties, e.g., sulfites in wine and vegetable oils with added anti-oxidants.

Group 3 Foods

The items in this group are simply called processed foods. Typically, they are derived by combining Group 2 foods with Group 1 foods, e.g., additions of sugar, oil, and salt. Processed foods utilize various preservation techniques, cooking methods, and fermentations, as in breads and cheese. Two main features of Group 3 foods are (1) to maintain the integrity of the foods via preservation using methods like canning, bottling, salting, sweetening, curing, smoking, etc. and (2) to enhance flavor. Also, specific preservatives to prevent microbial activity may be used. Some beverages, such as beer and cider (derived from Group 1 foods), are included in this category.

Group 4 Foods

This last group is the key one as regards the degree of processing. It serves as the motivating factor for most of the research studies reviewed in this chapter. The items in this group include ultra-processed foods and drinks. As noted earlier, ultra-processed foods are industrially sourced products; they typically contain five or more ingredients. They may include ingredients from Group 3. However, they also include ingredients not found in any of the other groups, such as artificial and natural flavors and colors, additives extracted from natural foods (e.g., casein, lactose, whey) or synthetically produced substances (e.g. hydrogenated oils, hydrolyzed proteins, protein isolates, maltodextrin, high fructose corn syrup, etc.), stabilizers, non-sugar sweeteners (natural and artificial), and various processing aids (bulking agents, defoamers, anti-caking agents, emulsifiers, sequestrants, etc.). Rarely will Group 1 foods show up in Group 4. Some constituents of Group 4 require special industrial techniques (e.g., extrusion, distillation, centrifugation, etc.). Typically, foods in this group are (1) designed to be consumed with minimal preparation or no preparation, (2) hyperpalatable (exceptionally tasty), (3) packaged in branded and showy containers, (4) aggressively marketed, and (5) owned by national and international companies. Tens of thousands of items fall into this group: soft drinks, snacks, ice cream, mass-produced baked goods, candies, industrial oils, margarines and spreads, cereals, energy bars, fruit drinks, infant formulas, most foods in the frozen section of the grocery store, distilled liquors (e.g., rum, whisky, gin) and so many more. In fact, one could make the claim that the majority of foods found in mega supermarkets and convenience stores in the USA would be classified as Group 4 foods.

Since the advent of the NOVA food classification system, many researchers world-wide have used it to evaluate associations between diet and nutrient intake, obesity, metabolic syndrome, abnormal blood lipid (fat) levels, consumption of added sugar, cardiovascular disease, and other diseases. Also, the NOVA system has been influential in the development of dietary guidelines in some countries.

Before moving on to the research studies, I need to pause here and say something about the NOVA classification system compared to the Processed Food Index (PFI) that I devised to evaluate fast foods. The two systems have both similarities and differences.

The NOVA system is simpler with only 4 categories for ingredients vs. 5 categories for the PFI system. Group 1(unprocessed) in NOVA corresponds to the "unprocessed" category in the PFI system. Group 2 (culinary) loosely corresponds to the "lightly processed" category. Group 3 (processed) corresponds to "moderately processed" and "highly processed" categories combined. Group 4 (ultra-processed) roughly corresponds to the "extremely processed" category. The major deviation between the two systems arises with the purpose of the Processed Food Index, which attempts to quantify the degree of industrialization of a fast-food menu item based on the descriptors of the individual ingredients composing the food. The NOVA system does not take the additional step of labeling specific industrialized foods (e.g., a fast-food chain pizza pie) with a numerical measure of the degree of processing.

Further Readings:
"NOVA. The Star Shines Bright"

World Nutrition Journal, Jan-Mar 2016, Monteiro, CA, et. al.

https://www.worldnutritionjournal.org/index.php/wn/article/view/5/4

The Nature of Scientific Studies and Published Research Papers

In this chapter, I want to review research studies that shed some light on the relationship between the consumption of ultra-processed foods and impacts on human health. But, before going there, I need to acknowledge that readers have a diverse understanding of how research studies are conducted, how data is interpreted and conclusions drawn, and the language of science.

Scientists have a variety of research designs available to them to answer a scientific question. Some are more valid than others. Some work better than others for a given question. For a review of research designs, I will be drawing on the paper entitled "Overview of Clinical Research Design" by Daniel Hartung and Daniel Touchette published in 2009.

Let's pose this question: Are diets high in ultra-processed foods unhealthy for humans?

There are two general categories of research: experimental and observational. An investigator of an experimental research design could gather a group of people and, over a period of time, control their intake of ultra-processed foods and measure an effect(s) (outcome) of eating that type of diet. The best type of an experimental design is a random controlled trial (RCT). In that study, a group of people, say 100 individuals, would be selected. The total group would be split into two sub-groups of 50 by random selection. One group (the control group) would be placed on a typical diet with normal amounts of ultra-processed foods, while the other group (experimental) would be provided a diet high in ultra-processed foods. Over a period of time, let's say 10 weeks, each group would be monitored for specific health measures, e.g., blood pressure, cholesterol, blood glucose, etc. The best RCT studies are single blind and double-blind ones. In a single blind study, the subjects don't know which group (control or experimental) they are assigned to. In a double-blind study, both the subjects and the investigators don't know who has been assigned to which group. Blinded studies minimize biases that can affect the validity of study conclusions. At the end of the experimental study, the data would be collated, evaluated, and statistically analyzed for significance. Conclusions could then be made to determine whether the experimental group consuming the most ultra-processed foods was measurably unhealthier than the control group.

When it comes to diet research, the RCT study is very difficult to carry out for the following reasons: (1) In order to assure compliance, the participants would require isolation in a clinical setting, such that their meals could be strictly controlled; (2) The length of time needed to see differences (effects) between groups would be difficult to know; (3) If a long study time is needed, say a year, isolating the two groups in a food laboratory for that length of time would be impractical, unethical, expensive, and possibly cruel. For those reasons, RCTs, involving questions of dietary effects, are rarely conducted. However, later in this chapter, I'll review one study, an exception to the rule, called a cross-over RCT.

Note that over about 50 years ago, there were little or no restrictions on conducting rigorous control studies involving diet. A famous example was the 1944 starvation study conducted by Ansel Keys, a physiologist at the University of Minnesota. The U.S. government was concerned that huge numbers of starving people at the end of WWII would offer an enormous challenge to relief efforts, plus, America could gain world-wide prestige if it was prepared to feed the world's starving people during a period of famine. Keys gathered a group of 36 college-aged men (conscientious objectors to war) to participate in a study where the participants ate controlled diets in three phases: (1) first 3 months of normal feeding with exercise to establish a baseline of health and fitness, (2) then 6 months of starvation on a bland diet designed to imitate the diets of war victims (e.g., macaroni, turnips, weak soup, etc.), and (3) finally 3 months of controlled re-feeding to increase the amount of food and calorie intake during recovery. The men were required to walk 22 miles per week. The mental health of the men was followed as closely as their bodies. Some men experienced severe weight loss (e.g., 24% reduction) and suffered from emotional distress. In 1950, Ansel Keys published a two-volume report of the study entitled "The Biology of Human Starvation," a very expensive purchase at the Amazon bookstore.

The majority of dietary studies involve observational designs. These research experiments come in several flavors. The simplest and the least reliable observational design is the case study. Typically, a single individual is followed over time to determine how their health has changed. Maybe you've seen the 2004 documentary, "Super Size Me," in which the director, Morgan Spurlock, experimented on himself to see whether a diet dedicated to the consumption of McDonald's fast food would negatively impact his health. Conclusions from a sample population of one person can never be generally valid since a single individual can never represent a large population. At best, a case study might suggest cause and effect, which could lead to a larger, more controlled study with a bigger group of participants to verify an effect.

The best observational designs are cohort studies. A cohort is a group of individuals participating in a study. Those individuals are followed over time to determine if their personal diet gives rise to effects that study investigators are interested in. There are two types of cohort studies: prospective and retrospective. In a prospective study, individuals are followed in real time for a specified period. For example, a group of 100 people are asked to keep a food diary over a year's time. The investigators regularly query the participants regarding their food intakes, follow changes in their medical conditions during the study period, and then determine if there are any health outcomes associated with their diets.

In a retrospective study, a data set collected in the past is evaluated to determine if there were associations between diet and health outcomes, which could then be extrapolated into the present. Many epidemiological studies (medical research seeking to find the causes of diseases and disorders) use retrospective studies. For example, looking at past medical records showing exposure to arsenic in food could be related to the number of deaths over time from arsenic poisoning.

Both types of observational designs have their advantages and disadvantages. In both cases, the participants don't have to be isolated in a food laboratory and have their diets restricted. Studies may be conducted over a long period of time with large sample populations. For prospective studies, selection bias is limited because investigators cannot know ahead of time which participants will exhibit effects in the future. Selection bias, a type of experimental error, occurs when study participants are inappropriately chosen based on a high probability of exhibiting an anticipated outcome. Downsides of a prospective study include (1) the expense, since they take a long time to complete and (2) information bias may occur. Information bias, another experimental error, occurs when the data collected is not accurate. In the situation where participants are self-reporting about their diet, inaccuracies could creep in due to poor memory (recall bias) or other reasons. On the other hand, retrospective studies, since they use existing data, are less expensive to conduct and don't rely on the direct input of participants in real time.

Most scientific research papers address experimental errors. The quantity and quality of errors in an experimental study will obviously affect the validity of the conclusions from the study. Every study exhibits random errors (chance), which results from random variation in the data. A more likely source of error arises from bias in the data or confounding errors. Two sources of bias, selection and information, were mentioned above. A third source of bias is systematic error. This type of error occurs when there is a consistent inaccuracy in the data set. For example, a faulty analytical instrument consistently gives an erroneous result which is either consistently high or consistently low giving rise to an invalid data set. In a research study, random errors and biases are measured using statistical tests and techniques. The results of statistical examinations of data shed light on the validity of scientific conclusions. Researchers wanting to maximize the validity of their study conclusions will seek to minimize errors of all kinds. Since statistics is a language of its own involving complicated, mathematical algorithms, discussion of this aspect of experimental design is beyond the scope of this book. The research studies selected for discussion below have been published in major scientific journals and vetted by authorities in the field of epidemiology, so I won't be commenting on the statistical elements of the papers. However, before leaving the subject, I do want to comment on the remaining source of error: confounding errors. Simply put, these errors interfere with the interpretation of an experimental effect (outcome). The problem is best illustrated by an example. In studying the adverse effects on health by the consumption of a diet high in ultra-processed foods, what if many of the study participants lived in food deserts where ultra-processed foods were more plentiful and cheaper? That bias might falsely result in higher disease outcomes relative to a more balanced sample population. The results of the study would then be in question. Researchers attempt to minimize confounders by adjusting for them, using more representative population samples, and introducing randomization whenever possible.

Some scientists and science writers think that epidemiology studies, such as the prospective studies discussed in this chapter, are not valid, since they rely on the self-reporting of study participants and they are full of confounders. George Zaidan, a science communicator and television host, who is currently the executive producer for the American Chemical Society, challenges the conclusions drawn from the associations found in prospective studies. Here is a quote from his 2020 book "Ingredients: The Strange Chemistry of What We Put in US and on Us," a zany, sometimes vulgar look at how science works. He examines some of the same research studies regarding ultra-processed foods that I cover below. Here is a quote from Zaidan's book.

> "As I write this, most of the evidence against ultra-processed foods is observational, and the risks are not huge, which makes it low-quality. Low-quality evidence is the kind of evidence that you look at and go, 'Huh, that's interesting. Maybe we should do a randomized controlled trial to test if this association is legit and causal.' It's *not* the kind of evidence that you look at and go, 'We've conclusively concluded with 100 percent 'conclusivity' that this association is legit and causal. Alert the media!' To be clear, all I'm saying here is that we should be cautious about the evidence on ultra-processed food. I'm *definitely* not saying that eating processed food is *good* for you … Ultra-processed food could turn out to be an important cause of obesity and diabetes, but it could also turn out to be mostly a passenger variable, tapping on the death accelerator a little, but not the major driving force. We don't know yet. Research dollars will continue to be spent on studying ultra-processed foods, so the evidence will eventually get better. Who knows: one day it might even be good enough to start imprisoning bags of Cheetos. But that'll take years."
> [Italics applied by George Zaidan.]

Further Readings:

Overview of Clinical Research Design

Daniel Hartung and Daniel Touchette

https://tinyurl.com/exupmtxm

Super Size Me

2004 Documentary

https://tinyurl.com/2d7t4wc5

The Great Starvation Experiment

Todd Tucker

https://tinyurl.com/2yqjxvbe

The Biology of Human Starvation

Ansel Keys, et. al.

https://tinyurl.com/257xdgm7

Ingredients: The Strange Chemistry of What We Put in Us and On Us

George Zaidan, Dutton, 2020

https://tinyurl.com/26jqgx3x

Lastly, for the nonscientist, reading and understanding scientific papers can be a major challenge, because every science discipline has its own language. Not knowing the terminology and techniques in a scientific field can pose a significant roadblock to even getting through the abstract of a paper, much less reading the whole paper. In summarizing articles in the remainder of this chapter, I have tried to simplify the scientific language into layman's terms, so readers can understand the gist of the experimental design and the outcomes of the research.

Ten Scientific Papers Addressing Health Issues Associated with Ultra-Processed Foods

#1: NutriNet-Santé Study - Cancer

The first study to examine comes out of France. Its umbrella name is the NutriNet-Santé study. Out of this mammoth study came several sub-studies which will be mentioned later. Since this study is the granddaddy of ultra-processed food research projects, I will go into some detail about its structure. Subsequent reviews of other studies will not be covered in as much detail.

In May 2019, the British Journal of Medicine published an article entitled, "Consumption of Ultra-processed Foods and Cancer Risk: Result from NutriNet-Santé Prospective Cohort." The investigators only focused on the associations between diets high in UPFs and cancer, which is the second leading cause of death globally ... about 1 in 6 deaths are due to cancer. Notice that the research project is described as a "prospective cohort," which tells us that the study was an observational one and conducted in real time. The NutriNet-Santé study was designed to observe a large group of people over a long time period to determine if there was a measurable relationship between a diet high in UPFs and the risk of developing cancer. What makes this a mammoth study was the size of the cohort and the length of time that participants were tracked. The cohort size was 104,980 people (aged 18 and above recruited via multimedia campaigns) who were followed for about 9 years from 2009 to 2017. The initial cohort was even larger by about 10,000, but volunteers got eliminated who had histories of serious diseases or who didn't follow study guidelines. The overall dropout rate was 6.7%.

Each subject (participant) was asked to fill out three nonconsecutive, on-line, 24-hour dietary surveys involving 3300 items (foods and drinks). In the first two years, the investigators just monitored food intake to establish a baseline, then, in the remaining years, the subjects were monitored for incidents of cancer. To gather more information, subjects also filled out 5 questionnaires related to demographics (age, gender, income, etc.) and lifestyle attributes (dietary intakes, physical activity, personal/family health histories, etc.). The dietary surveys were issued over three non-consecutive days every six months and randomly assigned over a two-week period including 2 weekdays and 1 weekend day.

The first two years of data were averaged per individual to serve as the baseline of dietary intakes. The surveys were validated by trained dietitians and by use of blood and urinary markers. The portion sizes of all foods and beverages were reported. Subjects who under-reported or maintained low food intakes were booted from the study.

Dietitians categorized foods and beverages into one of four food groups according to the NOVA food classification system. Subjects reported major health events through yearly health questionnaires and received physical checkups every three months. Also, subjects were invited to submit their medical records and, sometimes, their doctors were contacted for additional information and verification.

The monitored cancer events included overall cancer occurrence, breast cancer, prostate cancer, and colorectal cancer. Each time a participant reported a cancer diagnosis, a follow-up was made with their physician and/or hospital to gather more information. Additionally, an expert committee of doctors reviewed the data. The outside medical information minimized potential bias of the subject's reports. Participants who died during the course of the study were followed up through a French death registry to determine causes of death.

Statistical models were used to assure the validity of the data. To minimize bias, these models were adjusted for age, gender, body mass index (BMI), height, physical activity, smoking status, alcohol intake, calorie intake, family history, educational level, nutritional factors, and others.

The participants who consumed the largest quantities of ultra-processed foods shared the following characteristics: younger age, current smoker, less educated, less family history of cancer, lower physical activity, higher caloric intake, consumed more fat, carbs, sodium, but lower alcohol intake. There were no significant differences in dietary intake between men and women (UPF intake in men and women was 17.6% and 17.3% respectively).

So, after crunching the data, what did the investigators find? For every 10% increase in the amount of UPF consumed, there was an associated risk of 12% in overall cancer occurrence and an associated risk of 11% in breast cancer in particular. They did not find any associated risk for prostate or colorectal cancers, although the latter did lean in that direction; however, the data could not be statistically validated. Interpreted a slightly different way, if study participant "A" ate just 10% more UPFs than study participant "B," then participant "A" had a 12% greater risk of getting cancer over the 9-year study. Things could get worse as the UPF consumption increased. If participant "A," ate 20% more UPFs, then their risk of getting cancer would double to 24%.

Getting down into the weeds of the study, the investigators found that ultra-processed fats (e.g., hydrogenated oils) and sugary food and beverage products were the big contributors to the risk for overall cancer and ultra-processed sugary products were strongly associated with the risk for breast cancer. As for the UPFs themselves, the largest proportions came from sugary foods, highly processed fruits and vegetables, beverages, starchy foods and cereals, and processed meats and fish.

Every observational study has its pros and cons. On the pro side: (1) Since the study cohort was very large and the time period was long, the chance of errors affecting the conclusions was relatively small; and (2) The study involved rigorous assessments of the dietary intakes of the participants, which boosted the accuracy of the data collected. On the con side: (1) The study was weighted heavily towards female participants, who tend to exhibit more health-conscious behaviors, plus they had higher educational levels and professional standing. Those characteristics set them apart from the average French citizen. This means that the investigators probably would have observed more deaths from cancer if there was a more representative sample population, and that would have produced a larger effect (higher percentage risks); (2) If there were NOVA misclassifications of UPFs (e.g., a food labeled as a UPF but actually belonged in another class or vice versa), then the data would be skewed by this systematic error, and (3) Some cancers are initiated decades before signs can be observed, so it's conceivable that some of the reported cancers in the study originated before the start of the study, again skewing the data.

We must realize that the best conclusions of an observational study can only refer to associations between the study variable (in this case consumption of UPFs) and the study effects (cancers). An observational study cannot prove a relationship because all other variables cannot be controlled for. That's why the double-blind random controlled study is the gold standard for proof. Of course, even those types of studies can be plagued by error, so proof is relative.

Consequently, the authors of the paper are able to show an effect, but not pin down the cause. However, they were able to propose some hypotheses to explain the association between high consumption of UPFs and an increase in cancer rates. Here are the main hypotheses:

- Diets high in UPFs are nutritionally deficient due to the higher amounts of calories, sugar, fat, and sodium, and they typically are low in fiber content and micronutrients. The constituents of UPFs (refined foods, additives, etc.) displace healthy ingredients found in unprocessed foods. The investigators actually found this cause to contribute the least risk of increased cancer rates.

- UPF-rich diets provoke higher glycemic responses (more glucose released in the blood) and lower satiety (feeling full and not overstuffing).

- People who consume diets high in UPFs tend to have higher body weights. The medical literature definitely links obesity with risk of cancer.

- Many synthetic chemicals are used in UPFs. Some of them have known carcinogenic effects (e.g., sodium nitrite in processed meats and acrylamide generated in baked and fried starchy foods).

- Scientists have not studied the possible cumulative and interactive adverse effects of additives in UPFs. Multiple additives ingested over time may trigger cancer events.

- The packaging material used for UPFs may harbor contaminants like bisphenol A, an endocrinal disrupter, linked to cancer.

Further Readings:
Consumption of Ultra-processed Foods and Cancer Risk: Results from NutriNet-SantéProspective Cohort

British Medical Journal, February 2018

Thibault Fiolet,, Bernard Srour, et.al. (Including Carlos A. Monteiro)

https://www.bmj.com/content/360/bmj.k322.long

#2: NutriNet–Santé Study – Cardiovascular Disease

Using the database from the French NutriNet-Santé study discussed above, investigators examined the relationship between diets high in UPFs and the development of cardiovascular diseases. This study, entitled "Ultra-processed Food Intake and Risk of Cardiovascular Disease: Prospective Cohort Study," was published in the British Journal of Medicine in May 2019.

Heart disease, the leading cause of deaths, is responsible for one-third of all deaths globally. The investigators looked at heart disease events that included stroke, transient ischemic attack (mini-stroke), myocardial infarction (heart attack), acute coronary syndrome (clot in coronary artery), and angioplasty (medical procedure to open coronary arteries).

Here are the results of the study:

- The risk of cardiovascular diseases (CVD) increased across all levels of the sample population irrespective of gender, age, fat consumption, body mass index (BMI), and degree of physical activity.

- For every 10% increase in the percentage of UPFs in the diet, there was an associated 12%, 13%, and 11% statistically significant increase in the rates of overall cardiovascular, coronary heart, and cerebrovascular disease (stroke), respectively. For example, if you were a study participant being compared to a sub-group who didn't consume any UPFs and 10% of the foods you ate were UPFs, then you had a 12% higher risk of having a cardiovascular event by the end of the study. If your UPF consumption was 20% higher, then the risk went up by 24%.

- Conversely, the results of the study indicated that there was a significantly lower risk of developing cardiovascular disease if participants consumed a diet high in unprocessed or minimally processed foods.

- Even after data adjustments for nutritional issues associated with UPFs in the diet, there was still an increased risk for CVD, suggesting that there were components of UPFs directly affecting the development of heart disease.

- Speculation on the cause of the increased risk of CVD included higher consumption of calories, sodium, fat, and sugar and lower consumption of fiber.

- Higher intakes of fruits and vegetables were protective against developing CVD.

- Displacement of nutrients by UPFs in the diet created nutritional deficits.

- Some additives are triggers of CVD such as sulfite (preservative), monosodium glutamate (flavor enhancer), carboxymethylcellulose (thickening agent), polysorbate-80 (emulsifier), carrageenan (thickening agent), and acesulfame K (artificial sweetener).

Further Readings:
Ultra-processed Food Intake and Risk of Cardiovascular Disease: Prospective Cohort Study

British Journal of Medicine, May 2019

Bernard Srour,, Léopold Fezeu, et.al. (Including Carlos A. Monteiro)

https://www.bmj.com/content/365/bmj.l1451

#3: NutriNet-Santé Study – Mortality

Using the database from the French NutriNet-Santé study, investigators examined the relationship between diets high in UPFs and an increase in overall mortality (death) risk. This study, entitled "Association Between Ultraprocessed Food Consumption and Risk of Mortality Among Middle-aged Adults in France," was published in the Journal of the American Medical Association in February 2019.

For this study, deaths of participants were verified with the French national registry. To assess mortality during the course of the study, only participants 45 years or older were followed because that sub-cohort was considered most likely to succumb to nutrition-associated causes of death. Also, participants who died during the baseline period were excluded from the analysis. That left a sample population of 44,551 people. This group consisted of 32,549 women (73%) and 12,002 men (27%). UPFs on average accounted for 14.4% of the foods consumed and 29.1% of the total calories. Intakes of UPFs increased with older people (45 to 64 years), lower income, lower education level, living alone, higher BMI, and lower physical activity. During the study period there were 602 deaths including 219 from cancer and 34 from CVD. The results showed that after adjusting for confounders, a 10% increase in the amount of UPFs consumed was associated with a 14% higher risk of all-cause mortality.

Here are some hypotheses offered by the authors of the paper to explain the correlation:

- UPFs often are high in sodium content which is associated with CVD and stomach cancer risk.

- UPFs often are high in added sugar which is also associated with CVD.

- Processed meats, prevalent UPFs in western diets, have been associated with CVD.

- Given that UPFs are usually low in fiber, adverse health effects can result from that nutritional deficiency.

- The combination of high caloric intake with high intake of fat, sugar, and salt is linked to the development of noncommunicable diseases and increased mortality.

- Increased exposure to toxic chemicals in UPFs can increase mortality risk.

- High dietary intakes of UPFs can lead to metabolic diseases which increase mortality risk.

Further Readings:
Association Between Ultraprocessed Food Consumption and Risk of Mortality Among Middle-aged Adults in France

Journal of the American Medical Association, February 2019

Laure Schnabel, Emmanuel Kesse-Guyot, et.al.

https://jamanetwork.com/journals/jamainternalmedicine/fullarticle/2723626

#4: Association Between Consumption of Ultra-processed Foods and All-Cause Mortality: SUN Prospective Cohort Study

It's interesting to note that just a year after the initial results of the French NutriNet-Santé Prospective Cohort were published in the British Journal of Medicine, a similar cohort study, based in Spain, was published in 2019 in the same journal. Instead of specific health issues, like cancer, this study focused on all-cause mortality, which looked for associations between UPF consumption and all causes of death.

The study was conducted by a consortium of researchers associated with the University of Navarra in Pamplona, Spain. The acronym SUN refers to Sequimiento Universidad de Navarra (Follow-up at the University of Navarra). The UPF findings were just a subset of this very large study. From December 1999 to February 2014, an initial sample population of 22,114 university graduates were surveyed for food and drink consumption. The food and drinks were classified according to the NOVA system. The survey consisted of a food frequency questionnaire consisting of 136 items. The study was weighted towards women (12,113). The age range was 20 to 91 years. During the 15-year study, participants were surveyed every two years. Only 165 participants were excluded from the study due to not meeting a total minimum or maximum daily calorie intake and 2215 people were dropped from the study because they did not meet the two-year survey requirement (the retention rate was 90%). The final cohort size was 19,899.

A total of 335 deaths occurred during the study period. UPF consumption was categorized into quarters (low, low-medium, medium-high, and high). The results showed that a high consumption of UPFs (>4 servings daily) was associated with a 62% increased risk of death from all-cause mortality. For every additional serving of UPF, the all-cause mortality increased by 18%.

Just to bring it back to reality, here are a few of the UPFs that were monitored: pudding, ice cream, ham, all processed meats, pate, foie-gras, chips, cereals, pizza, margarine, baked goods, chocolates and candies, carbonated beverages, fruit drinks, milkshakes, instant soups, mayonnaise, and distilled liquors.

It's no surprise that the sub-group of participants with the highest consumption of UPFs had higher, average BMIs, were most likely to smoke, and had a family history of chronic diseases. Also, this sub-group was more likely to snack, watch television, nap, consumed higher amounts of fat, ate more fast foods, and had lower consumptions of fruits and vegetables.

An obvious criticism of this study was that the cohort was not representative of the Spanish population, since it was a relatively young population and was restricted to university graduates. As a consequence, the number of deaths over the study period was lower than normal. Also, the socioeconomic status of the sample population was probably higher than that of the general population.

Further Readings:

Association between consumption of ultra-processed foods and all cause mortality: SUN prospective cohort study

BMJ 2019;365:l1949, May 2019

Anaïs Rico-Campà, Miguel A Martínez-González, et.al

https://www.bmj.com/content/365/bmj.l1949

#5: Association of Ultra-Processed Food Consumption with Increased Risk of All-Cause and Cardiovascular Mortality: The Moli-sani Study

In parallel with the French and Spanish studies, a similar investigation was conducted in Italy. The results were published in December 2020. This research was a collaboration of several medical institutions and universities in Italy. In examining the name of the study (Moli-sani), "Moli" refers to the region of southern Italy (Molise) where the study was conducted and "sani" refers to health. From 2005 to 2010, 22,475 men and women (> 35 years of age) from the Molise region were recruited with a participation period of 8.2 years. The average age of the participants was 55.

As in the earlier studies that were discussed, study participants were asked to report food intakes using questionnaires. Food items were categorized using the NOVA classification system. The intake of UPFs ranged from under 6.6% to over 14.6% of the total diet. The median intake of UPFs per day was 0.4 lbs. The bulk of the UPFs came from processed meats, pizza, and cakes/pies.

Individuals reporting the highest intakes of UPFs were found to have 58% increased risk of cardiovascular disease mortality (death), 52% increased risk of ischemic heart disease (IHD) (narrowed heart arteries) and cerebrovascular disease (CVD) (stroke), and 26% higher risk of death from all causes. Interestingly, no association with large consumptions of UPFs was found with cancer deaths. The study found a strong association between high sugar consumption and deaths from IHD/CVD.

Further Readings:
Ultraprocessed Food Consumption Is Associated with Increased Risk of All-Cause and Cardiovascular Mortality in the Mole-sane Study

American Journal of Clinical Nutrition, 2021;113:446–455.

Marialaura Bonaccio, Augusto Di Castelnuovo, et.al.

https://tinyurl.com/232tuque

#6: Consumption of Ultra-processed Foods and the Impact on Canadian's Health

In this Canadian study, rather than look at UPF consumption associated with mortality over many years, researchers took a snapshot of Canadian's diet during a specific year, 2001, looking at the proportion of calories coming from UPFs based on grocery store purchases. Of course, the researchers were analyzing data from over 20 years ago, so how trustworthy is it today? However, given the nature of the North American food industry, the availability and consumption of unhealthy processed foods has significantly increased since 2001 (see Science Daily reference below). If this study was repeated today, the results would be even more skewed towards diets high in UPF consumption.

The data on food consumption was obtained from the Canadian Food Expenditure Survey (FOODEX). The survey provides estimates on quantities of foods and drinks purchased by households across Canada (98% coverage). Sub-samples were taken for this study that focused on households of similar socio-economic standing. A total of 5643 Canadian households were surveyed. Over a 14-day period, participants reported food purchases including weight amounts. Food purchases at restaurants were not included. Interviewers visited participating households to assure that diaries were properly completed. Then the researchers classified the foods using descriptions based on a modified NOVA system. Group 1 foods included unprocessed or minimally processed products; Group 2 included processed culinary ingredients; and Group 3 included UPFs. The mean daily calorie consumption per household member was calculated, which turned out to be 2129 Cal/day.

Then the total calories were broken down into the three food groups. The results showed that 61.7% of total calories came from UPFs, 12.7% came from Group 2 foods, and 25.6% came from Group 3 foods. The overall diets were examined for the amount of fat, saturated fat, free sugars, and sodium and were found to exceed the levels recommended by the World Health Organization. Also, the diets were lower in fiber content. The researchers observed that "only 20% [of] of the lowest consumers of ultra-processed products ... were anywhere near reaching all nutrient goals for the prevention of obesity and chronic, non-communicable diseases." To change that trend "would mean a fundamental change, from a reliance on ready-to-eat or ready-to-heat Group 3 products to preparation and cooking of meals based on Group1 foods and Group 2 ingredients."

Further Readings:

Consumption of Ultra-processed Foods and Likely Impact on Human Health. Evidence from Canada

Cambridge University Press, November 2012

Jean-Claude Moubarac, Ana Paula Bortoletto Marings, et.al

https://tinyurl.com/28wwfqba

Americans Are Eating More Ultra-processed Foods: 18-Year Study Measures Increase in Industrially Manufactured Foods That May Be Contributing to Obesity and Other Diseases

Science Daily, October 2021

New York University

https://tinyurl.com/yfhjvklf

#7: Consumption of UPFs and Cardiovascular Disease: USA Framington Offspring Study

Back in 1948, the US government sponsored an observational study called the Framingham Heart Study (FHS), the purpose of which was to shed light on the causes of heart disease. The population of Framingham, Massachusetts, near Boston, was selected for participation in the study. The initial population sample was 5,209 men and women ranging in age from 32 to 60. The examination and interpretation of data over decades gave rise to what we know today as the risk factors for heart disease, including smoking, family patterns, high blood pressure, high cholesterol levels, unhealthy eating, lack of physical activity, and obesity. Since 1948, participants have been surveyed every 2 and 6 years including the collection of detailed medical histories and lab testing.

In 1971, the study was continued with the second generation of Framingham citizens. The latest study, initiated in 2002, continued with the grandchildren of the original cohort and is called the USA

Framingham Offspring Study. There have been over 3,000 peer-reviewed scientific papers published based on the FHS data.

In the Framingham Offspring Study, researchers investigated associations between UPFs and cardiovascular (CVD) incidence and mortality. The cohort included 3,003 adults who initially were free of CVD. Data was collected via food frequency questionnaires, anthropometric measures (weight, height, BMI, etc.), and sociodemographic and lifestyle factors over the period 1991 to 2008. Clinical examinations were obtained every 4 years. Mortality data were collected until 2017. As in studies discussed earlier this chapter, the UPFs were defined according to the NOVA system. Out of the 3,003 participants (mean age 53), 1,062 experienced incidents of CVD. On average, study participants consumed 7.5 daily servings of UPFs at baseline.

Statistical analysis of the data showed that, with <u>each</u> additional serving of UPFs, 7% of the sample experienced an increase in hard CVD, 5% experienced an increase in hard congenital heart disease (CHD), and 9% experienced in increase in CVD mortality. Hard CVD or CHD is characterized by a buildup of plaque in coronary arteries. There was no association between the consumption of UPFs and total mortality. The authors of the paper suggested that the limiting UPFs in the diet would decrease the occurrence of CVD.

Further Readings:

Ultra-processed Foods and Incident Cardiovascular Disease in the Framington Offspring Study

Journal of the American College of Cardiology. 2021 Mar, 77 (12) 1520–1531

Filippa Juul, Georgeta Vaidean, et.al.

https://www.jacc.org/doi/full/10.1016/j.jacc.2021.01.047

Framingham Heart Study website

https://www.framinghamheartstudy.org/participants/

#8: Weight Gain from Diets High in Ultra-processed Foods - USA

So far, the research studies highlighted in this chapter are observational studies. An observational study, as regards discovering associations between diet and adverse health effects, requires participants to fill out food diaries or frequency questionnaires pertaining to their eating patterns. The obvious drawbacks to this approach in collecting data involve the integrity of each participant (are they following the rules?), the reliance on the participant's memory (how many of us can remember what we ate yesterday?), and the tendency of some people to skew data to make themselves look good (who doesn't have an ego that can act up occasionally?). Researchers investigating diet-related diseases certainly realize these shortcomings. That's why observational studies usually have large cohorts (multi-thousands of participants) and extend over long periods of time (5+ years). In examining the large amount of data collected over the course of a study, researchers can often weed out bad or incomplete data. If bad or inaccurate data gets through the screening process, the usual small number of bad data points gets swamped by the massive amount of good data, so the effect of the errors is small when the complete data set is interpreted. Also, using complex, state-of-the-art statistics programs helps in cleaning up the data.

In this USA project, instead of an observational study, a random controlled trial (RCT) of diet was conducted. As mentioned earlier, RCTs are the highest tier in scientific studies since they are independent of participant bias. Typically, two groups of participants are randomly selected with the groups being matched as best as possible (e.g., by age range, dietary pattern, medical history, gender, etc.). The members of each group are "blind" as regards what diet they are consuming. Ideally, the administrators of the study are also "blind," so they don't know which diet each participant is consuming during the study. In this way, overall bias is kept to a minimum. In an RCT study of this type, it's paramount that the diet of each participant is controlled, i.e., the participants agree to reside in a clinical setting for a certain period of time where they are fed very specific foods according to the assigned diets. Although RCTs are the gold standard of scientific studies, it's easy to see why they are not so common in food research projects: (1) Participants must agree to withdraw themselves from their normal lives and (2) It's expensive to set up a food laboratory/kitchen with strict controls over food preparation and meals served over the requisite period of the study. It's no surprise that these projects have small cohorts and are much shorter in length relative to observational studies.

This study involved 20 adults (average age 31, BMI of 27). There were 10 males and 10 females. The subjects were admitted to a metabolic ward of the National Institute of Health (NIH) Clinical Center for 28 days. They were randomly assigned to either an ultra-processed or unprocessed diet over two weeks after which the two groups switched diets. This is an example of a cross-over study. In this version of an RCT, bias is further reduced by having participants experience the same diet but at different times. The selected diets followed the NOVA classification system. The diets varied substantially. The ultraprocessed diet was much higher in the ratio of added sugar to total sugar (~54% vs 1%), insoluble fiber to total fiber (~16% vs 77%), and saturated to total fat (~35% to 19%) For both diets, meals were designed to provide the same number of calories, macronutrients (carbs, fat, and protein), sugar, sodium, and fiber. Interestingly, the subjects could consume as much or as little (ad libitum) as they wanted. Three meals per day plus snacks were offered. Study subjects maintained similar exercise routines during each diet period.

The researchers found that people on the ultra-processed diet consumed 508 more calories on average, mostly coming from carbs and fat. Weight gain accompanied the increase in food consumption. An average of 2 lbs was gained over the 2-week period. In contrast, the same amount of weight was lost while consuming the unprocessed diet. As reported by the participants, in terms of "pleasantness" and "familiarity," there were no significant differences between the two diets. Why the weight gain with the UPF diet? Simply put, the foods with higher UPF content may have just been more palatable causing people to eat more of them. The study concluded that (1) UPFs facilitate overeating and (2) limiting the consumption of UPFs may be an effective way to lose weight and prevent/treat obesity.

One criticism of the study's results points out that participants consumed food in a controlled environment much different from their normal environments, which would include many more impacts on their eating patterns (e.g., stress, job demands, family issues, etc.). As a consequence, the results of the NIH study could be skewed.

Further Readings:
"Ultra-Processed Diets Cause Excess Calorie Intake and Weight Gain: An Inpatient Randomized Controlled Trial of Ad Libitum Food Intake"
Clinical and Translational Report, Volume 30, Issue 1, Pages 67-77.e3, July 02, 2019
Kevin D. Hall, Alexis Ayuketa, et.al.
https://tinyurl.com/y9jrrsrn

#9: Highly Processed Foods and Addictive-Like Eating Behaviors - USA

All the studies discussed so far have dealt with chronic diseases and mortality, but a critical issue that many Americans are dealing with is food addiction. Do UPFs influence that kind of behavior?

In 2015, two psychologists and a neuroscientist published an article detailing an investigation into what processed foods were associated with addictive-like eating behaviors. Actually, two studies were conducted. In Study One, 120 undergraduates from the University of Michigan completed the Yale Food Addiction Scale (YFAS) and then selected, out of 35 foods of varying nutritional composition, which ones were associated with addictive-like eating behaviors. In Study Two, 384 participants were recruited through MTurk, an Amazon platform for hiring temporary employees. In this study, the researchers investigated which food attributes (e.g., fat grams) were associated with addictive-like eating patterns. Then they determined the individual differences for that association.

Here is what they found. Although UPFs are not specifically mentioned in the article, reference is made to highly processed foods. In Study One, they found that processed foods, high in fat and glycemic load (GL), were most often associated with addictive-like eating patterns. What is glycemic load? It's a measure of carbohydrate intake as it affects blood sugar. To define GL, another measure is needed: glycemic index (GI). The GI is a number from 0 to 100 assigned to a food. The 100 value corresponds to pure glucose, a reference food. For any other food, the rise of blood glucose in 2 hours is measured against pure glucose. Foods with high GIs (70 or higher) release glucose to the blood quickly and significantly, e.g., white rice, white bread, corn flakes, and peeled white potatoes. Foods with medium GIs (56 to 69) include sugar, whole wheat bread, banana, and sweet potatoes. Foods with low GIs (55 or less) include beans, lentils, nuts and seeds, unprocessed breads, and most fruits. To calculate a glycemic load of 100g of a food involves multiplying its carbohydrate content in grams by its GI value and dividing by 100. For example, watermelon has a GI value of 72. A 100g serving of watermelon has 5g of carbohydrates. Multiplying 72 by 5 and dividing by 100 gives 3.6. By comparison, a 100g serving of white bread, with a GI of 70 and carbohydrate amount of 54.2g, would yield a GL of 37.9, i.e., 70 times 54.2 divided by 100. The white bread provides a significantly higher glycemic load (about 10 times higher). A diet designed to control glycemic load tries to avoid sustained blood-sugar spikes, and it can help type 2 diabetics manage their blood sugar.

In Study Two, the degree of processing was associated with degrees of food addiction. The researchers concluded that not all foods played a role in addictive behavior, but highly processed foods, which, like drugs of abuse, appeared to be strongly associated with food addiction." The shared characteristics included persistent desire, unsuccessful attempts to quit, personal activities (work, social events, etc.) were given up or reduced, usage continued despite knowledge of adverse consequences, and tolerances developed requiring increased usage with lower rewards. In the author's words, "… in our modern food environment, there has been a steep increase in the availability of what is often referred to as 'highly processed foods', or foods that are engineered in a way that increases the amount of refined carbohydrates (i.e., sugar, white flour) and/or fat in the food … It is plausible that like drugs of abuse, these highly processed foods may be more likely to trigger addictive-like biological and behavioral responses due to their unnaturally high levels of reward … Additionally, addictive substances are altered to increase the rate at which the addictive agent is absorbed into the bloodstream. For example, when a coca leaf is chewed, it is considered to have little addictive potential. However, once it is processed into a concentrated dose with rapid delivery into the system, it becomes cocaine, which is highly addictive. Similarly, highly processed foods, compared to naturally occurring foods, are more likely to induce a blood sugar spike … it appears that highly processed foods may be altered in a manner similar to addictive substances to increase the food's potency (dose) and absorption rate."

In conclusion, the researchers specifically mentioned refined carbohydrates, fats, and their combination in highly processed foods as triggers for a high glycemic load. The level of processing indicated a strong association with addictive-like eating behavior.

Further Readings:
Which Foods May Be Addictive? The Roles of Processing, Fat Content, and Glycemic Load

PLoS One (Public Library of Science); 10(2): e0117959

February 18, 2015

Erica M. Schulte, Nicole M. Avena, et.al.

https://www.ncbi.nlm.nih.gov/pmc/articles/PMC4334652/

#10: How UPFs Promote Chronic Diseases Across a Number of Research Studies – USA

Sometimes the conclusions of a research project can be amplified by combining the results of similar studies in what is called a meta-analysis. The authors of this paper performed a systematic review of all the observational studies that investigated the relationship of UPFs and health status. The review included 23 studies across multiple countries (10 cross-sectional and 13 prospective cohort studies).

Here are the results for the cross-sectional studies involving 113,753 participants. High UPF consumption was associated with:

- Increased risk of being overweight or obese (+39%).

- Increased weight circumference (+39%).

- Reduced high-density lipoproteins (HDL) cholesterol (+102%).

- Increased metabolic syndrome (+79%).

The percents in parentheses refer to increased risks in experiencing the health effects.

No issues with hypertension (high blood pressure), hyperglycemia (high blood glucose), or hypertriacyglcerolemia (high fat level in blood) were observed.

Here are the results for the prospective cohort studies with a grand total of 183,491 participants who were surveyed for a period ranging from 3.5 to 19 years. High UPF consumption was associated with:

- Five studies showed 25% increased risk of all-cause mortality.

- Three studies showed 29% increased risk in cardiovascular disease.

- Two studies showed 34% increased risk in cerebrovascular disease.

- Two studies showed 20% increased risk in depression.

In conclusion, the authors of this study postulated an association between high UPF consumption and a worse cardiometabolic risk profile.

Further Readings:
Consumption of Ultra-processed Foods and Health Status: A Systematic Review and Meta-analysis
British Journal of Nutrition; 2021 Feb 14; 125(3): 308–318.
G. Pagliai, M. Dinu, et.al.

https://www.ncbi.nlm.nih.gov/pmc/articles/PMC7844609/

CHAPTER 16

INTERNATIONAL SCORING
SYSTEMS

"Forget about nuclear war; the real weapons of mass destruction are fast food, highly processed carbs, and high-fructose corn syrup!"

— Brian Quebbemann, M.D.,
bariatric surgeon

Now, let's pause a moment and review the topics covered so far. An analysis of the ingredients in the menu offerings for three highly popular fast-food restaurants (McDonald's, Pizza Hut, and Taco Bell) revealed that a large percentage of the ingredients are ultra-processed. In turn, many of the individual menu items could be classified as highly or extremely industrialized. Next, the scientific literature was surveyed looking for studies that associated high consumptions of ultra-processed foods with deleterious health outcomes and mortality. Consistent findings among studies from a variety of countries showed that these types of foods adversely affect the health of consumers, particularly in the long term. Given the link between chronic, lifestyle diseases (heart disease, stroke, obesity, diabetes, metabolic diseases, etc.) and the consumption of industrialized foods, are governments doing anything about it? Are consumers being warned? Are food manufacturers and chain restaurants required to label products to inform their consumers of the relative health hazards associated with eating foods packed with ultra-processed ingredients? Those questions need answers. In this section, I survey a several countries attempting to introduce scoring systems for processed foods. I begin with the United States, my home country, and proceed to others.

Food Scoring in the United States

Sadly, in one of the wealthiest and most scientifically advanced nations of the world, there is no federal scoring system for processed foods. U.S. citizens are left to fend for themselves when it comes to evaluating the health hazards of highly processed foods. If people can decode the Nutrition Facts Label on packaged foods and identify highly processed ingredients, they stand a fighting chance, but, seriously, most people won't bother with this evaluation because it's complex and time-consuming, and many don't have the educational background to do a good job of it. So, what's a citizen to do? Well, obviously, they can't count on the federal government for assistance, but there are alternatives, which will be discussed at the end of this section.

I don't mean to imply that the U.S. government doesn't look out for the health of Americans. There is a long history of the federal government passing legislation to protect consumers against food adulteration and poisonings, misrepresentations on packages, false claims, the use of hazardous ingredients, and so on. One government agency particularly stands out in educating citizens in adopting a healthy diet. That's the United States Department of Agriculture (USDA) founded in 1862 during Abraham Lincoln's first administration. Now, if you think about it, why would an agency in the executive branch of the government responsible for overseeing, regulating, and promoting agricultural production in the country be mandated to help citizens eat healthier? After all, most of the ultra-processed ingredients found in the standard American diet (SAD) are derived from agricultural crops and livestock products promoted by the USDA. What gives? Let's look at the mission statement for the USDA.

We provide leadership on food, agriculture, natural resources, rural development, nutrition, and related issues based on public policy, the best available science, and effective management. We have a vision to provide economic opportunity through innovation, helping rural America to thrive; to promote agriculture production that better nourishes Americans while also helping feed others throughout the world; and to preserve our Nation's natural resources through conservation, restored forests, improved watersheds, and healthy private working lands.

The USDA provides multiple services under its jurisdiction to deliver its commitment to the American public, including: Farm Service Agency, Natural Resources Conservation, Food & Nutrition, Food Safety & Inspection, Agricultural Marketing, Animal & Plant Health Inspection, Forest, Rural Development, Agricultural Research, National Institute of Food & Agriculture, Foreign Agriculture, and U.S. Codex Office.

It's clear from the USDA mission statement that this single federal department juggles both the promotion/support of agricultural production and food nutrition/food safety. How can one department simultaneously advocate for farmers planting and selling commodity crops like corn, soy, and wheat that populate, directly and indirectly, some of the unhealthiest foods sold in grocery stores and restaurants, while, at the same time, guide consumers towards the healthiest diets possible? Do we have a conflict of interest here? We'll see some examples of this conflict later.

Let's focus on the dietary guidance that the USDA has provided by looking at a short, historical review of the department's advocacy for the health of the American people during the last 120+ years.

1894: USDA publishes the "Farmer's Bulletin" which included a few food recommendations for the populace.

1916 to 1930s: Two publications are issued called "Food for Young Children" and "How to Select Food" (for adults). These booklets guided consumers on food group purchases and household measures. Food groups such as milk, meat, cereals, vegetables, fruits, fats and fatty foods, as well as sugary foods were mentioned.

1933: USDA introduces food plans at four different cost points to advise people during the Great Depression

1941: The first Recommended Dietary Allowances (RDAs) were created that listed specific daily intakes of calories, protein, calcium, and vitamins.

1940s: "A Guide to Good Eating" is published highlighting the Basic Seven food groups including (1) green/yellow vegetables, (2) oranges, tomatoes, grapefruit, (3) potatoes and other root vegetables, (4) milk and milk products, (5) meat, poultry, fish and eggs (oddly also included beans, peas, nuts, and peanut butter), (6) bread, cereals, and flours, and (7) butter and fortified margarine (added Vitamin A). No specific serving sizes were recommended.

1956 to 1970s: USDA published "Food for Fitness, a Daily Food Guide." The Basic Four food groups were emphasized including (1) vegetables and fruits, (2) milk, (3) meat plus beans, peas, and peanut butter, and (4) cereals and breads (including pasta and rice). Note that specific amounts of foods were recommended, but there was no mention of the specific intakes of fats, sugars, or calories.

1970s: The USDA turns its attention to research that focuses on the overconsumption of certain foods that contain fat and cholesterol. The research results were criticized by the meat and dairy industries. The government finally responds to the criticism in 1979.

1977: The "Dietary Goals" for the United States was released. Senator George McGovern's Select Committee on Nutrition and Human Needs suggested that the USDA Dietary Goals include this statement, "Consume only as much energy as you expend, eat more naturally occurring sugars, consume more fruits and vegetables and go easy on eggs and butter." These guidelines were vociferously opposed by the scientific establishment and the food industry citing lack of sufficient scientific evidence.

1979: USDA publishes "Hassle-Free Daily Food Guide." This booklet was released after the 1977 Dietary Goals was released and, in addition to the basic four food groups, and, in a nod to the meat and dairy industries, included a fifth food group that highlighted the moderate intake of fats, sweets, and alcohol.

1980: USDA releases the first comprehensive guidelines for Americans called "Nutrition and Your Health: Dietary Guidelines for Americans."

1984: The new dietary guidelines were entitled "Food Wheel: A Pattern for Daily Food Choices." The USDA turns to graphics to get their message across. Five food groups were mentioned along with recommended amounts at 3 different calorie levels.

1992: The next dietary guidelines introduced the now iconic Food Guide Pyramid, which was in use through 2011.

2005: The publication "MyPyramid Food Guidance System" accompanied the next release of the Dietary Guidelines for Americans (DGA). A band was added to the pyramid to include oils (another nod to the soy and corn industries?). Also, physical activity gets a mention. A 10-page consumer brochure was issued called "Finding Your Way to a Healthier

You" to provide advice to consumers about smart food choices.

2011: USDA introduces a new graphic "MyPlate" introduced with the 2010 DGA.

2014: The Agricultural Act of 2014 (Farm Bill) mandates that by 2020 the DGA include dietary guidance for infants and toddlers, as well as women who are pregnant.

2020: The 9th and current version (as of 2022) of the DGA is published.

Here are some details about how the Dietary Guidelines for Americans are created. Congress mandated, starting in 1980, that a new set of guidelines was to be released every 5 years to update the contents with respect to new nutritional and medical research. Two federal departments were to take turns spearheading the effort: the U.S. Department of Agriculture (USDA) and Health and Human Services (HHS). The effort would be a 3-step process: (1) The agency involved would identify specific nutrition topics to be reviewed; (2) An external Dietary Guidelines Advisory Committee (DGAC) was to be appointed comprised of expert medical doctors, nutritionists, dietitians, medical researchers and other health professionals to review the latest scientific evidence related to the selected topics; the report issued by the DGAC would be reviewed by a committee of bureaucrats who would then collectively put together the next issue of the DGA. **Note:** A surprisingly small group of experts were appointed to the DGAC. They were considered temporary federal employees who would serve for two years. In the first few decades, fewer than 10 people were on the committee. For the 2020 DGA, there were 20 members on the committee. Still, to think that only 20 people are making recommendations for the whole country somewhat boggles the mind.

Since 1980, controversy has swirled around the DGAs as competing interests fought to get their opinions heard and to influence the final publications. Just imagine how huge beverage companies, like Coca-Cola and Pepsi, would react to recommendations by the DGAC to reduce sugar in the American diet. They would likely take a big hit to their bottom line, so there were significant lobbying efforts by individual food & beverage companies or their trade organizations (e.g., American Beverage Association, Consumer Brands Association, and Corn Refiners Association). Money and politics don't mix well when it comes to the interests of the average American. At the same time, various organizations advocated for significant improvements in the DGAs to safeguard people's health. Additionally, each DGA review period allowed public comments and hearings were held to receive even more input. Many people, businesses, and organizations wanted their shot at influencing the final outcome.

Marion Nestle, a consumer advocate, has followed the battles over the DGAs since their inception. She is the author of many books including "Food Politics." She currently has a blog where she provides frequent commentary and criticism on the DGA process and other food-related topics. I interviewed Ms. Nestle in January, 2021 for the "Food Labels Revealed Podcast" (see reference below)." She is a very enthusiastic and feisty defender of consumer rights. What follows are some excerpts from her blog regarding the 2020 DGA.

> *"The report, by the way, is 295 pages. Do we really need all this? The guidelines stay pretty much the same from edition to edition: eat more fruits and vegetables (plant foods); don't eat much salt, sugar, saturated fat; maintain healthy weight. Or, as Michael Pollan famously put it, 'Eat food. Not too much. Mostly plants.' The food industry has the biggest stake in dietary guidelines, which is why we have to go through all this." (8/18/22)*

"The wording changes from edition to edition. The editions get longer and longer. And the basic problems—nutrients as euphemisms for the foods that contain them, more and more obfuscation—stay the same ... What is the relationship between consumption of dietary patterns with varying amounts of ultra-processed foods and growth, size, body composition, risk of overweight and obesity, and weight loss and maintenance? Comment: This was one of my big criticisms of the 2020-2025 guidelines; the word "ultraprocessed" was never mentioned, yet I consider it the most important nutrition concept to come along in decades. So, this is a big step forward ... What is the relationship between beverage consumption (beverage patterns, dairy milk and milk alternatives, 100% juice, low- or no-calorie sweetened beverages, sugar-sweetened beverages, coffee, tea, water) and growth, size, body composition, risk of overweight and obesity, and weight loss and maintenance? Risk of type 2 diabetes? ... What is the relationship between food sources of saturated fat consumed and risk of cardiovascular disease?" (4/19/22)

"My strongest criticism of the 2020 dietary guidelines is that they fail to say anything about the health benefits of reducing consumption of ultra-processed foods (the junk food category strongly associated with excessive calorie intake, weight gain, and poor health). Yet here we have a published review in a food science journal arguing for debunking 'myths' about food processing. They are not myths. Evidence is abundant." (11/23/21)

"Salt: The report says remarkably little about sodium beyond that it is overconsumed and people should "reduce sodium intake." It's possible that I missed it, but I could not find suggestions for quantitative limits. Ultra-processed: The word does not appear in the report except in the references. A large body of evidence supports an association of ultraprocessed foods to poor health. If the committee considered this evidence, it did not spell it out explicitly." (7/16/20)

Have you heard about the "revolving door" in Washington D.C.? The term refers to business people who seek jobs in regulatory agencies of the federal government and vice versa. It's a derogatory term implying that business leaders bring personal biases into their new jobs as bureaucrats, who can then unfairly influence decisions made in government to favor the businesses that they came from. It works both ways. Former bureaucrats are often offered lucrative positions in business, who then bring their government expertise with them to advise businesses how to exert influence on government legislation or rule making. This "revolving door" movement of personnel between government and business can lead to unhealthy relationships between the private sector and regulating bodies.

A great example of the "revolving door" problem occurred in August of 2017. Kailee Tkacz, a former director of food policy with the Corn Refiners Association, a trade group representing corn syrup manufacturers, applied for a job with the USDA to assist in developing the DGA. Despite the conflict of interest as spelled out in the Ethics Pledge, the White House counsel, Donald McGahn, issued a waiver for Ms. Tkacz. In earlier jobs, Ms. Tkacz also lobbied for the Snack Food Association and the National Grocers Association. Previously, lobbying the USDA, the Congress, and other federal agencies, her new job allowed her to help set policy, which could directly benefit the industries with which she was formerly associated.

Another controversial concern in the development of DGAs is the conflict of interests (COIs) associated with the selection of the members of the Advisory Committee. A paper published online by Cambridge University Press in March 2022 shed some shocking light on that issue. Investigators researched the connections between Advisory Committee members and the food industry. Such connections included research funding by the industry; editing of a publication run by a business; hiring as speakers for paid events; participating in an organization of industry sponsored events; serving as board members on advisory/scientific committees that have liaisons with industry; becoming an employee or consultant for a food business; or receiving prizes or awards provided by the food industry. Here are the findings from the paper:

"Our analysis found that 95% of the committee members had COI with the food, and/or pharmaceutical industries and that particular actors, including Kellogg, Abbott, Kraft, Mead Johnson, General Mills, Dannon, and the International Life Sciences had connections with multiple members. Research funding and membership of an advisory/executive board jointly accounted for more than 60% of the total number of COI documented."

The paper's authors noted strong connections between committee members, and major companies. Sometimes, the prospective Advisory Committee members were nominated by major food companies.

The authors of the paper issued a guideline for future selections to serve on the DGAC:

"For Americans to be able to trust the guidance from the DGA as sound, objective, and science-based, it is imperative to ensure that each step of the process, from the selection and appointment of the DGAC to the final release of the DGA, is publicly accessible, transparently administered, and largely free of COI and influence from actors whose profit-driven interests are often at odds with those in public health. Our analysis of COI of DGAC members has shown that this is far from true."

One of the DGAC members, Sharon Donovan, PhD, RD at the University of Illinois, a large land-grant university in Urbana, IL with close ties to agriculture industries, had 152 conflicts of interest with 31 industry players.

Let's get back to the main subject of this section, namely how the United States government is dealing with the correlations between diets high in ultra-processed foods and serious health problems. You can get a big clue by downloading the PDF of the DGA (see reference at end of section). If you perform a key word search on "processed food" or "ultra-processed food" or "junk food" or "fast food," you will find **zero** hits, although 510 hits showed up with the word "healthy."

The answer to the questions "Is the USDA and HHS addressing the overconsumption of ultra-processed foods by Americans?" the obvious answer is emphatically, "No." Needless to say, there's no legislation or policy dealing with a scoring system for processed foods. Despite the epidemic of obesity and lifestyle diseases tied to food consumption, the U.S. federal government is doing squat to address significant causes. Later in this section, we'll see that government oversight of packaged foods varies according to country.

Here's an FDA update from September 2022, just a few months before the publication of this book. The FDA has proposed a rule that updates the 1994 definition of "healthy" based on the nutrient content of a food. The original, very conservative definition required foods to provide 10% of the Daily Value for one or more of vitamin A, vitamin C, calcium, iron, protein, and fiber before food manufacturers could use the word "healthy" on package labeling. Also, independently, the FDA is developing a front-of-package label that would designate a commercial food as healthy. Not surprisingly, there is no government mandate to designate a food as healthy or unhealthy. As far as food producers are concerned, the FDA guideline would only be voluntary.

Let's take a closer look at the FDA recommendations. Note that raw, whole fruits and vegetables would automatically qualify for the "healthy" claim based on their nutrient profiles. Here are some examples of foods that could be deemed healthy if the dietary requirements are met. [DV: Daily Value]

Food Group	Equivalent Serving	Added Sugar Limit	Sodium Limit	Saturated Fat Limit
Grains	¾ oz whole-grain	5% DV (2.5g)	10% DV (230mg)	5% DV (1g)
Dairy	¾ cup	5% DV (2.5g)	10% DV (230mg)	10% DV (2g)
Vegetable	½ cup	0% DV	10% DV (230mg)	5% DV (1g)
Fruits	½ cup	0% DV	10% DV (230mg)	5% DV (1g)

Seafood	1 oz	0% DV	10% DV	10% DV
Egg	1	0% DV	10% DV	10% DV
Beans	1 oz	0% DV	10%	5%
Nuts & Seeds	1 oz	0% DV	10%	5%
100% Oil	N/A	0% DV	0%	20%
Dressing, oil-based (30% oil)	N/A	2% DV	5% DV	20%

Source: Author Generated Table based on the FDA proposed rule for "healthy" foods.

Products that currently qualify as 'healthy' that could not be labeled as such under the FDA proposed definition would include white bread, highly sweetened yogurt, and highly sweetened cereal.

Certainly, with this new proposal for healthy foods, the U.S. government is heading in the right direction, but it falls short of the more stringent labeling by other countries. The fact that the proposed labeling is voluntary and there is no evaluation of the degree of processing are severe limitations of the proposed rule.

***

Despite the lack of intervention and guidance by the U.S. government, there are some attempts to characterize processed foods by the private sector. Two examples of non-commercial organizations are described below.

The non-profit Environmental Working Group (EWG) has a food scoring system called "Healthy Living." It is accessible by digital devices and the internet.

Image Source: EWG website
https://www.ewg.org/foodscores/
[Note: The color graphic at the website shows a range of colors from left to right ranging from green to yellow to orange to red.]

Here are a few example foods:

Go to the webpage: https://www.ewg.org/foodscores/

In the search field at the top right of the webpage, type in Fritos, then perform a search. When the products page appears, click on "Fritos the Original Corn Chips." The product shows a score of 4.0. "The score is based on weighted scores for nutrition, ingredient, and processing concerns. Generally, nutrition counts most, ingredient concerns next and degree of processing least. The weighted scores are added together to determine the final score." The best scores are close to 1 while the worst are close to 10. In this case, Frito chips have a decent score despite the fact that they are highly processed. What helps this product is that there are only 3 ingredients in it and none are artificial. You can see the contributions of nutrition (N), ingredient (I), and processing (P) by looking at the bar scale on the webpage. In this case, the Frito chips get rough scores of N=3.3, I=0.5, and P=0.2. Also on the same page, you'll see Nutrition Facts info similar to what appears on the package. I think this is a very good scoring system and provides a considerable amount of information. The system can search over 80,000 foods, which is impressive.

Let's try another example. Type in bologna and search. Select Armour Chicken Bologna. The overall score is 8.0 … not healthy. The ingredient score is of highest concern. If you scroll down the page and look at the left side, EWG lists all the negatives and positives regarding this food. There are 8 negatives (e.g., carcinogenic ingredient) and 1 positive (high protein). Look at the ingredient list on the left side of the webpage … lots of industrial ingredients!

Let's look at a healthier food. Type in Mom's Cereal. Select Mom's Best Cereals Toasted Wheatfuls. The overall score is 1.6 … very healthy. This cereal only has 2 ingredients (one is a vitamin), it's made from a whole grain, and there are plenty of positive findings.

The one criticism that I would make about the EWG scoring system is that the algorithm used is not transparent. It doesn't look like EWG wants to reveal the food scientist behind the curtain who generates the final food scores.

The EWG scoring system does a good job of combining the key factors for food quality: nutritional value, types of ingredients, and degree of processing. If adopted by the U.S. government, it could provide a suitable labeling scheme for processed food packaging as a means to inform consumers about the quality of the foods they eat.

***

A second example of a scoring system is the Food Compass, a nutrient profiling system created by a team of scientists at the Friedman School of Nutrition Science and Policy at Tufts University in Massachusetts. This tool is intended to help consumers, food companies, restaurants, and cafeterias choose and produce healthier foods and to assist agencies in developing sound public nutrition policy.

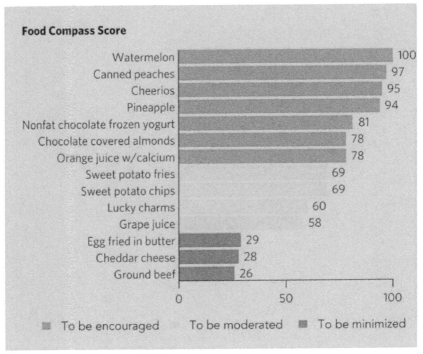

Food Compass Score

Watermelon	100
Canned peaches	97
Cheerios	95
Pineapple	94
Nonfat chocolate frozen yogurt	81
Chocolate covered almonds	78
Orange juice w/calcium	78
Sweet potato fries	69
Sweet potato chips	69
Lucky charms	60
Grape juice	58
Egg fried in butter	29
Cheddar cheese	28
Ground beef	26

■ To be encouraged　　To be moderated　■ To be minimized

Image Source: Food Lies Facebook page
[Note: The color graphic at the Facebook page shows three colors from top to bottom ranging from green (above 69) to yellow (above 29) to red.]

The Food Compass system considers both healthful vs. harmful factors in foods. It attempts to include latest evidence from nutrition science, food ingredients, processing traits of food, phytochemicals, and food additives. The system is applicable to all foods and beverages, even mixed meals. It was developed and tested using a national database of 8,032 foods and beverages consumed by Americans. It's complex algorithm scores 54 different characteristics across 9 domains incorporating a wide array of health measures. The system is flexible enough to expand if additional food attributes become available in the future. The scoring system provides values from 1 (least healthy) to 100 (most healthy). Foods and beverages falling in the 31 to 69 range should be consumed in moderation, while foods scoring 30 or less should be minimally consumed. Across various food categories, the average score was 43.2. Typical values for types of foods are sweet desserts (16.4), vegetables (69.1), fruits (73.9), legumes, nuts, and seeds (78.6), starchy vegetables (43.2), beef (24.9), poultry (42.67), and seafood (67.0). However, there are some unexpected results. For example, Cheerios scores a 95, but corn flakes come out at 19 (probably due to the large difference in sugar content); sweet potato chips and whole-grain bulgur (often viewed as a health food) have the same rating of 69. Unhealthy foods correlate heavily with cholesterol levels. Plus, no distinction is made between nutrients that occur naturally in foods versus those that are artificially added.

Further Readings:
"The History of USDA Nutrition Guides"

https://www.manufacturing.net/home/blog/13190282/the-history-of-usda-nutrition-guides

"The U.S. Food Guidelines Have Always Been Controversial"

https://time.com/4125642/dietary-guidelines-history/

"Table: History of Dietary Guidance Development in the United States and the Dietary Guidelines for Americans – A Chronology"

https://tinyurl.com/2p8a5kac

"Dietary Guidelines for Americans 2020 – 2025"

https://tinyurl.com/2c8ry5f7

Marion Nestle blog

https://www.foodpolitics.com/

"Food Politics: *How the Food Industry Influences Nutrition and Health*" by Marion Nestle.

https://tinyurl.com/2x5myaak

Food Labels Revealed Podcast

"An Interview with Marion Nestle, "Food Labels Revealed Podcast," January 2021

https://foodlabelsrevealed.podbean.com/e/flr-058-an-interview-with-marion-nestle/

"Corn Syrup Lobbyist Is Helping Set USDA Dietary Guidelines"

https://tinyurl.com/3bhkfaer

"Conflicts of Interest for Members of the U.S. 2020 Dietary Guidelines Advisory Committee"

Melissa Mialon, Paulo Serodio, et.al.

Published online by Cambridge University Press: **21 March 2022**

https://tinyurl.com/2p89tvj8

"FDA Proposes to Update Definition for "Healthy" Claim on Food Labels

U.S. Food and Drug Administration (FDA)"

https://tinyurl.com/yx64e7yk

"Use of the Term Healthy on Food Labeling"

U.S. Food and Drug Administration (FDA)"

https://tinyurl.com/25e5rku7

Environmental Working Group (EWG)

https://www.ewg.org/foodscores/

Healthy Living App

Tufts University

https://sites.tufts.edu/foodcompass/

Food Compass

Food Scoring Systems in Other Countries

Having established that the United States government has yet to establish a standard food scoring system for front-of-package processed foods, the obvious question is what are other countries doing to inform consumers about the quality of foods that they are eating? Are other countries stepping up to curb the unhealthy trends of obesity and chronic diseases? The answer is a definite "YES." I can't reasonably cover all the countries of the world in this chapter, but I will cover the most popular and effective scoring systems starting with European countries.

European Countries

The Nutri-Score System

This scoring system for commercial foods has been implemented in the last 5 years, but it originated back in 2005. Mike Rayner, a professor in the Department of Public Health at the University of Oxford, developed a calculation for rating the healthiness of manufactured foods based upon their negative and positive impacts on health. There was a negative impact tally (N) influenced by such factors as high caloric content, high sugar content, high amount of saturated fat, and high salt content. The positive impact tally (P) was influenced by the presence of fruits, vegetables, nuts, legumes, fiber content, protein content, and the presence of healthier oils (e.g., olive and walnut). When the P tally was subtracted from the (N) tally, a nutritional score was determined.

Interestingly, the French government adopted the Raynor system in March 2017, which was to be displayed on food products as a Nutri-Score. Apparently, Raynor's home country, Great Britain, was not interested in adopting the Raynor system. Without going into all the details of the calculation, the final results gave values varying from -15 to +40. The <u>lower</u> the score, the higher the health rating. To make interpretation easier for consumers, the numbers were correlated with the letters A, B, C, D and E which are displayed on the front of food packages. The best score is A and the worst is E. Additionally, the letters are color coded to provide additional visual information. The main objective with this scoring system was to allow consumers to compare the overall nutritional values across products in the same category, e.g., cereals. With that informed comparison, consumers could decide to eat less of the lower ranked products and more of the higher ranked ones.

Number Range	Letter	Color
-15 to -1	A	Dark Green
0 to 2	B	Light Green
3 to 10	C	Yellow
11 to 18	D	Orange
19 to 40	E	Red

Ranking of commercial foods using the Nutri-Score system. Note that only the last two columns show up on the label on the front of the packaged food.
Source: Author Generated Table

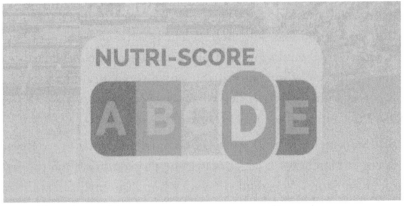

Image Source: Food Watch.org
[Note: The color graphic at the Facebook page shows a range of colors from left to right ranging from green to orange to red.]

At this time, the Nutri-Score system is voluntary, i.e., food manufacturers are not required to display the Nutri-Score label. In addition to France's adoption, Belgium, Spain, Switzerland, Germany, Luxembourg, and the Netherlands are using this scoring system. Some countries, such as Poland, Italy, and Greece are opposed to the Nutri-Score system citing that it does not give fair treatment to all foods. Internationally, this system is used in the Open Food Facts search engine to be discussed later.

What are the drawbacks and criticisms? There have been no large scientific studies showing that the Nutri-Score system has a positive outcome as far as consumer health is concerned. The system is not intended to compare foods from different categories, so if you're comparing the health benefit of a frozen pizza vs. a breakfast cereal, the scores are not going to be very helpful in choosing one food over the other. The system does not address the amounts of food consumed. The degree of processing of a manufactured food is not taken into account, which is the tool I've used for evaluating fast foods.

In the references listed at the end of this section, you'll find a website which allows anyone to calculate a Nutri-Score based primarily on the Nutrition Facts Label. Here are the data for Original Frito Corn Chips:

Per 32 chips or 28g.

Calories =	160
Total Fat (g) =	10
Saturated Fat (g) =	1.5
Trans Fat (g) =	0
Cholesterol (mg) =	0
Sodium (mg) =	170
Total Carbs (g) =	16
Dietary Fiber (g) =	1
Total Sugars (g) =	0
Protein (g) =	2

After plugging in the numbers, a Nutri-Score of 18 is obtained. That corresponds to a high "D" or orange rating. Fritos is rated low for its nutritional value. Note that, when the same product was evaluated by the Environmental Working Group's scoring system (see above), it came out in the middle range (4.0 out of 10.0 with lower scores indicating a healthier product). This difference is probably due to the fact that the EWG score takes into account the degree of processing, which gives Fritos a leg up.

Further Readings:

Wikipedia: "Nutra-Score"

https://en.wikipedia.org/wiki/Nutri-Score

"Nutri-Score Calculator"

https://nutriscore.impag.ch/

Colruyt Group: *"Make Smarter Choices with the Nutri-Score"*

https://www.colruytgroup.com/en/conscious-consuming/nutri-score

United Kingdom

Since the United Kingdom is no longer a member of the European Union, they are doing their own thing. The "Traffic Light Label" was first proposed in 2006 for Front-of-Pack (FOP) labeling to support consumers in making smart decisions about the food products they purchase. It focuses only on nutritional profiles (e.g., fat, salt, carbs, and sugar) and excludes evaluation based on the degree of processing. Based on the traditional traffic light colors of red, amber, and green, the scores depend on daily intake levels. In 2013, the Traffic Light system was formally introduced to the citizens of the UK, and food companies could voluntarily use it or not. Although not mandatory, most retailers (about 67%) have chosen to adopt the system. If adopted, there are several regulations that must be followed to ensure compliance. The panel must include all macronutrients, calories (energy), and salt, and serving size is based on 100g or 100mL. The most favorable nutrient profiles would include more amber and green colors on the label. The percentages listed on the label refer to an adult's daily reference intake. For example, in the graphic below, saturated fat is listed at 30% meaning that the food consumed (grilled burger) will contribute to almost one-third of the daily recommended amount.

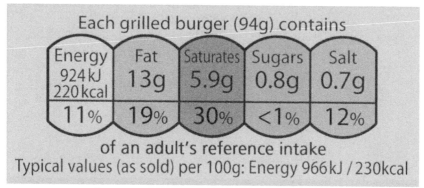

Image Source: Food Standards Scotland
https://www.foodstandards.gov.scot/business-and-industry/safety-and-regulation/labelling/nutrition-labelling-requirements

For this example, the colors across the panel from left to right are: white, amber, red, green, orange.

	LOW	MEDIUM	HIGH	
	Per 100g	Per 100g	Per 100g	Per portion
Fat	3.0g or less	3.0g - 17.5g	More than 17.5g	More than 21g
Saturates	1.5g or less	1.5g - 5.0g	More than 5.0g	More than 6.0g
(Total) Sugars	5.0g or less	5.0g - 22.5g	More than 22.5g	More than 27g
Salt	0.3g or less	0.3g - 1.5g	More than 1.5g	More than 1.8g

Image Source: University of Aberdeen, The Rowett Institute
The colors spanning low to medium to high are green, amber, and red.

Apart from informing consumers regarding the nutritional qualities of the foods they're eating, it is hoped that food manufacturers will reformulate their products to increase the nutritional benefits, i.e., showing more green and white colors than red and orange.

To evaluate Original Fritos Corn Chips as before would be difficult with the UK system since there is no final calculated score. However, using the nutritional facts for the product as listed earlier would allow a table to be created like the one below. From the table, a Traffic Light panel could be made based on 100g consumption.

	LOW	MEDIUM	HIGH
	Per 100g	Per 100g	Per 100g
Fat	-----	-----	35.7
Saturates	-----	-----	5.4
(Total) Sugars	0.0	-----	-----
Salt	-----	-----	1.5

Energy	Fat	Saturates	Sugars	Salt
571 kcal	36g	5.4g	0.0g	1.5g
29%	53%	27%	0%	26%

Source: Author Generated Tables

If colors were assigned to this Traffic Light table, the fat, saturates, and salt columns would be colored red and the sugars column would be green. The Fritos corn chips would not be recommended to consumers as a healthy food. This rating would be similar to the Nutri-Score, but not to the EWG rating, which also takes into account the degree of processing.

Further Readings:
"Five Reasons Why Multiple Traffic Light Front-of-Pack Nutrition Labelling Needs to be Made Mandatory in the UK"

https://tinyurl.com/2xhgvszo

"Guide to Traffic Light Labelling"

University of Aberdeen

https://www.abdn.ac.uk/rowett/documents/Traffic_light_guide.pdf

"Traffic Light Labeling UK – How to Apply to Your Food Label"

https://tinyurl.com/2dz5yh2g

"Check the Label"

https://www.food.gov.uk/safety-hygiene/check-the-label

Australia & New Zealand

Australia and New Zealand jointly have their own food scoring system called the Health Star Rating (HSR). This front-of-pack labelling system uses a star scorecard from ½ to 5 stars in ½ step increments. The more stars displayed, the healthier is the food item. The system was developed over a period of years starting in 2010 through a collaboration of federal, state and territorial governments with input from industry, public health, and consumer groups. The HSR was introduced to industry in 2014 with compliance on a voluntary basis. In March of 2018, a study found that about 10,300 products were displaying the HSR from more than 160 companies. On the downside, consumer groups criticize that companies are only displaying the logo on their best (healthiest) products, and not on the ones that would score low. Another criticism aimed at the HSR scoring system is that total sugars are taken into account rather than distinguishing natural sugars from added sugars. If the former were treated as good nutrients and the latter as bad nutrients, then some foods would get rated lower.

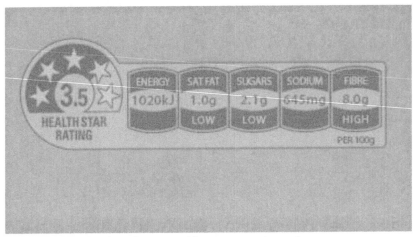

Image Source: Wikipedia webpage "Health Star Rating System"

Notice the features in the graphic above: (1) The number of stars both numerically and by the use of shading; (2) the energy content expressed as kilojoules (kJ) instead of calories; (3) the amount of saturated fat and a lower label indicating whether that value is low, medium, or high; (4) the quantity of sugars; (5) the amount of sodium, and (6) and the amount of fiber. All the nutrient amounts are based on 100g of food. Note that this scoring system is not color coded.

One of the drivers for the Health Star Rating system was the alarming growth of obesity rates in the citizens of both countries. In Australia, 63% of adults are overweight or obese plus one in four children have this medical problem. The proponents of HSR hope that with a simple graphic displayed on the front of food packaging, people will make better choices and reduce their calorie intake. Lastly, as with most of the other scoring systems discussed so far, the HSR does not take into account the degree of processing of a food item.

How many stars get displayed in the logo depends on the number of positive and risky nutrients in the food. The positive (healthy) nutrients include fiber, protein, and vegetable matter, while the risky (unhealthy) nutrients include fats and sugars. The mathematical component of the system is fairly complicated, but there is an automatic calculator available online, so you can determine an HSR for any food item if you have the nutrition facts available. The number that's calculated will only be an estimate because you'll also need the percentage of fruits and vegetables in the product, if any, and that information is generally not listed on packaged goods. If you want to play with the HSR, a link to the Excel calculator is listed in the references below.

Let's apply the HSR calculator to Original Fritos Corn Chips, the example used earlier for other scoring systems. The calculation gives a score of 2 stars. As expected, these corn chips don't warrant a healthy rating. No surprise there!

Further Readings:
"Applying the Health Star Ratings"

Health Star Rating - Applying the Health Star Ratings

"Health Star Rating System"

https://en.wikipedia.org/wiki/Health_Star_Rating_System

The Health Star Rating Calculator in Excel

https://tinyurl.com/253jkfau

Mexico

As of 2013, Mexico surpassed the United States as the most obese country in the world for adults (40% of population) and first for childhood obesity (26% of school-age children). By 2018, 75% of the adults were either overweight or obese. The government was keen on changing that trend by getting its citizens to pay more attention to the nutrient quality of processed foods.

Based upon the Mexican Daily Dietary Guidelines, a scoring system was approved in 2010 based on amounts of saturated fats, total fat, sodium, sugars, and energy content (calories) in foods. A front-of-pack label was devised showing the percentage of each nutritional measure per individual portions. In addition, the percentage of the recommended daily intake was displayed. After implementation of that scoring system, several studies indicated that the Mexican population was not responding to it. So, in 2016, the government revised the label basing it upon a simplified food labeling system created in Chile. By 2020, the new, non-voluntary labeling system was applied to 85% of food products.

The Mexican food scoring system is unlike any discussed so far. It does not involve any complex calculations. In fact, there is no actual score on the food label. The system consists of a set of seals that appear on the front of a food package that warn consumers about unhealthy components in the food. This is the simplest labeling system that I have seen so far.

Excessive saturated fats

Image Source: Wikipedia webpage "Food Labeling in Mexico"

The seal shown above, with a black octagon representing a stop sign, simply shows that the food product has an excessive amount of saturated fat per serving according to the Mexican dietary guidelines. There is no score. There are no numbers or percentages displayed. That information, as usual, would be in the nutritional content label somewhere else on the package. The complete set of seals address the following:

EXCESSIVE CALORIES

EXCESSIVE SUGAR

EXCESSIVE SODIUM

EXCESSIVE SATURATED FATS

EXCESSIVE TRANS FATS

CONTAINS (ARTIFICIAL) SWEETENERS. NOT RECOMMENDED FOR CHILDREN,

CONTAINS CAFFEINE, AVOID IN CHILDREN

The number of seals that wind up on the label would determine whether the food manufacturer would be allowed to offer incentives for purchase (rewards, toys, etc.), use pictures of celebrities, cartoon characters, etc. that would attract younger consumers. Additionally, if the package had any seals on it, the food manufacturer could not feature endorsements from medical organizations.

International suppliers of ultra-processed foods vehemently opposed the mandatory labeling in Mexico claiming that they were too restrictive (translated that means "Our sales will be adversely affected!") The huge beverage companies, Coca-Cola and PepsiCo, were among the opposition leaders. By 2021, studies showed that consumption of junky foods decreased but the sales levels did not. Some higher-minded companies have altered the nutrient content of their food and beverage products to reduce or eliminate the seals on their packaging.

Overall, the Mexican front-of-pack food labeling system is the most simplistic, but, unlike all other systems, the government mandates that all food manufacturers must use it. Unfortunately, this system does not address the degree of processing of industrial foods.

Further Readings:
Wikipedia, *"Food Labeling in Mexico"*

https://en.wikipedia.org/wiki/Food_labeling_in_Mexico

Chili

Around 2012, the Chilean government passed a law (No. 20606) dealing with the labeling of the nutritional composition of food products. A decree was issued in 2015 to implement the law. According to an earlier national health survey, 60% of Chile's population suffers from excessive weight. Looking at the combined weight issues for the country's children, over 50% of them are overweight or obese with many of them suffering from malnutrition. This problem of malnutrition is the main public health problem. The new law was an attempt to address unhealthy behavior patterns in the purchases of food. The law and its regulations came into full effect in 2016. Among other stipulations, (1) the law mandated the ministries of Education and Health to provide a nutritional monitoring systems for students at all grade levels; (2) the banning of sales of unhealthy foods in schools; (3) food companies would no longer be allowed to market or donate unhealthy foods to children under 14 as well as give away promotional items such as toys; (4) the banning of any baby formula labeling that discourages breastfeeding; and (5) specifically established limits for calories, sodium, sugars, and saturated fats in processed foods.

Image Source: Bloomberg.org

The above is an example of a label that could appear on the front of a packaged food sold in Chili. The stop-sign shaped logos warn that the food is high in calories, saturated fats, and sugars.

As we've seen, these types of food labels unfortunately ignore the degree of processing of high and ultra-processed foods. Hopefully, those warnings will be available in the future. The good news about the Chilean food law is the prohibition of unhealthy foods to school children and the prohibition of marketing them to children under 14 years of age. These standards are far more stringent than any proposal in the United States and other countries. A good example is soda, which is limited to 15g sugar per 8 ounces. For comparison, an 8oz Coke contains 26g of sugar in the USA.

A study by University of North Carolina researchers found that the purchases of beverages high in sugar decreased 23.7% in the years 2015 to 2017. They also examined beverages that were high in saturated fat, sodium, or calories.

Further Readings:
Chile – Urban Food Policy Snapshot

https://tinyurl.com/2cnypo3e

Brazil

The Brazilian food labeling law was updated in 2020 and rolled out in late 2022. It looks very similar to the Chilian regulations. For the first time, front-of-packaging rules were established. As usual, the changes were intended to help consumers make more informed choices regarding processed foods. There are also mandatory declarations on the nutrition facts label for (1) total and added sugars, (2) energy (e.g., calories) per 100g or 100 mL, and the number of servings per package.

Seals are required for front-of-packaging labels if particularly unhealthy components exceed designated limits in the food: (1) added sugars are greater or equal to 15g per 100g or 7.5g per 100mL; (2) saturated fat content greater or equal to 6g per 100g or 3g per 100mL; and (3) sodium content greater or equal to 600mg per 100g or 300mg per 100mL. Note that for an 8oz Coke, the added sugar limit would be about 17g, much lower than in the United States.

Image Source: mdpi.com, Multidisciplinary Digital Publishing Institute, a publisher of open access journals

The above label shows 6 different seals that could appear on a food package in Brazil if levels of sodium, sugar, total fat, saturated fat, trans fat, and sugars are exceeded. The adjacent amounts graphic provides additional details for a 25g serving. There is no required information for the degree of processing of the food.

It's incredibly odd that the Brazilian government, in creating the law regulating food companies as regards front-of-packaging labeling, ignored its native son and international authority, Dr. Carlos Monteiro, a major contributor to the Nutri-Score system.

Further Readings:
"Brazil: From October 2022 New Rules for Nutritional Labeling"
Merieux Nutrisciences

https://regulatory.mxns.com/en/brazil-new-rules-nutritional-labeling-october-2022

"Brazil: Nutrition Labeling: New Rules Take Effect in 120 Days"
Infoalimentario.com

https://tinyurl.com/2y2k5tkp

Other Countries

Of the countries not mentioned in this chapter, many of them have nutrition or dietary guidelines. According to the Food & Agricultural Organization (FAO) of the United Nations, among them are (1) seven countries in Africa have established dietary guidelines as of 2020 including South Africa, Nigeria, Kenya, and Namibia; (2) 18 countries in Asia and the Pacific as of 2020 including Afghanistan, Bangladesh, Cambodia, India, Indonesia, Japan, Malaysia, Philippines, Republic of Korea, Thailand, China, and Vietnam; and (3) 25 countries in Latin America and the Caribbean as of 2020 including Argentina, Bahamas, Columbia, Costa Rica, Cuba, El Salvador, Honduras, Jamaica, Panama, Paraguay, Peru, Uruguay, and Venezuela.

Further Readings:
"Food Based Dietary Guidelines"
Food & Agricultural Organization, United Nations

https://www.fao.org/nutrition/education/food-dietary-guidelines/background/en/

International Scoring System

In 2012, a very enterprising and intelligent French programmer by the name of Stéphane Gigandet launched a food scoring system called Open Food Facts, a free, online crowdsourced database of food products from all over the world. It's sometimes referred to as the Wikipedia of food scoring systems because anybody can contribute to the database. Since 2012, the database has been steadily growing and today it has over 2 1/2 million entries.

This non-profit project has a scoring system based upon the Nutri-Scores calculation, so that means, unlike most scoring systems, it incorporates the degree of processing of industrial foods. The Open Food Facts app is available online or can be installed in smart phones, so it's readily available to consumers all around the world. The app provides a boatload of information about a commercial product including ingredient lists, food additives, brand, and the nutritional value or score.

I tried looking for the score for our now familiar Original Frito Corn Chips using the app listed in the reference panel below (https://world.openfoodfacts.org/), but, disappointedly, I couldn't find that item. Let's instead look up Frito-Lay Corn Tortilla Chips. Near the top of the page, click on "most scanned products" using the dropdown list if necessary. Then from the dropdown list for "explore products by ...," select brands. Wait for the list to open. In the Search field, enter Frito-Lay. Scroll the rows of products until you find Corn Tortilla Chips. Click on the logo. Here are the findings.

Image Source: Open Food Facts Application

The Nutri-Score label in its color version shows, going from left to right: dark green (A), light green (B), dark yellow (C), orange (D), and red (E). That order represents healthy to unhealthy. The score of "D" for the corn tortilla chips is pretty low, not surprising for a corn chip product.

The Open Food Facts app shows even more data including (1) ingredients, (2) whether product is palm-oil free, vegan, or vegetarian, (3) high-risk additives, (4) nutrient levels per 100g, (5) and full nutrition facts.

Further Readings:
Open Food Facts - World

https://world.openfoodfacts.org/

Summary

In this analysis of food scoring systems around the world, there are a handful that have been introduced to consumers in specific countries to encourage wise food selections. Unfortunately, only the Nutri-Score system, in a subset of European countries, incorporates the degree of processing of industrial foods. As more and more studies are conducted showing the health issues around diets heavy in high and ultra-processed foods, maybe governments will seek to convey that information to their citizens. If they are really interested in improving public health, particularly in regards to obesity and lifestyle diseases, governments should be placing front-of-package labels on commercial foods that are easy to read and interpret.

I think both the Open Food Facts or the Environmental Working Group scores, converted to a front-of-packaging label, could be adopted by all countries since each conveys important nutritional information and takes into account the degree of processing. Instead of each country developing their own system costing time, effort, and money, a single universal system could be adopted and used by everyone on the planet. That makes perfect sense to me.

It's very sad to me that my home country, the United States, lags far behind other countries in encouraging or mandating front-of-package food labels that could help its citizens to make better decisions regarding healthy vs. harmful foods. Note that individual states, counties, and cities have made attempts (e.g., sugar limits in soft drinks) through legislation to reign in the scourge of injurious industrial foods, both those found in box stores and fast-food restaurants, but those isolated efforts have not had much of an effect on public health. Americans need a reliable and federally mandated labeling system to help them curtail their consumption of harmful foods.

CHAPTER 17

KEY TAKEAWAYS IN PART III

1. Thanks to advances in science, technology, and modern medicine, Americans are living longer than their ancestors, but their quality of life is diminishing, partly due to the massive consumption of industrial foods containing highly processed and ultra-processed foods.

2. Dr. Carlos Monteiro of Brazil, a leading researcher of the impact of ultra-processed foods on health, says that new synthetic foods, inventions of modern food technology, are not real foods and many are harmful to our health.

3. Public health officials fail to consider the effects of food processing and their link to obesity and chronic diseases, and focus mostly on the nutritional content of packaged foods.

4. Dr. Monteiro coined the word "ultra-processed," which refers to heavily marketed foods, high in additives making them more edible, palatable, and habit-forming.

5. Dr. Monteiro was instrumental in devising the NOVA system that categorizes foods according to the extent and purpose of food processing, rather than just in terms of their nutrient content; however, to date, the NOVA system has not been applied to fast food.

6. In recent years, many research studies have been conducted revealing the associations between diets high in ultra-processed foods and the development of chronic disorders

like obesity, metabolic syndrome, cardiovascular diseases, stroke, cancer, and others.

7. Nutrition and public health studies are epidemiological in nature, i.e., they seek to discover meaningful associations between disease and a given population.

8. Most research studies investigating the relationship of chronic disease and the degree of fast-food consumption use a prospective model in which a large population (cohort) is followed over a significant period of time. Participants record their food intakes in a systematic way. Scientists then collate the data, examine it for trends and patterns over time, and then report the results in a paper using statistical analysis to verify that the conclusions are valid. The best type of study, a random controlled study, is typically not useful for dietary research since it demands control over the behavior of people and over a long period of time, which is not realistic.

9. The NutriNet-Santé Study in France is the largest epidemiological study to date with food surveys of 104,980 people to determine correlations between consumption of ultra-processed foods and the development of cancer, cardiovascular disease, mortality, and other health conditions.

10. The United States has a long history of protecting consumers from contaminated foods, adulterated foods, and false marketing, as well as mandating that food companies create standard nutritional labels and ingredients lists.

11. Although better than nothing, the federal Dietary Guidelines for Americans are not determined strictly by the findings of nutritional science, but are selected by bureaucrats. The government officials may be directly or indirectly influenced

by players in the food industry through the "revolving door" in Washington D.C.

12. Probably due to the outsized influence of Big Food, the United States government lags woefully behind other countries in the world in advocating for front-of-pack labeling to promote healthy food choices for its citizens.

13. To fill the gap created by the United States government, several private organizations, such as the Environmental Working Group and the Friedman School of Nutrition Science and Policy at Tufts University (Food Compass project), have created food scoring systems for Americans that take both nutrients and degree of processing into account.

14. A number of countries have introduced food scoring systems to their populations. The most comprehensive and useful is the Nutri-Score system, which originated in France but has spread to other European countries. However, at this time, the requirement for posting labels is voluntary, not mandatory.

15. Brazil rolled out a mandated front-of-packaging labeling regulation in 2022, but, oddly, the government ignored its native son, Dr. Carlos Monteiro, who was a major influencer in the creation of the Nutri-Score system.

16. Most countries in the world have dietary guidelines, but very few address the issues surrounding highly or ultra-processed foods.

17. An open-source organization called Open Food Facts established a food scoring system in 2012 which has grown to contain over 2.5 million food items by 2022. This system is based on the Nutri-Score, so it addresses both nutrition and the aspects of food processing. Both this food scoring system and the one from the Environmental Working

Group, provides much better data than the ones adopted by world governments. Either one would serve as a good template to create a universal front-of-packaging labeling system to help people all over the world to improve the health of their diets and to stem the ever-rising tide of obesity.

PART IV

AND SO, IT ENDS

CHAPTER 18:

FINAL THOUGHTS

"Future historians, I hope, will consider the American fast food industry a relic of the twentieth century--a set of attitudes, systems, and beliefs that emerged from postwar southern California, that embodied its limitless faith in technology, that quickly spread across the globe, flourished briefly, and then receded, once its true costs became clear and its thinking became obsolete."

— Eric Schlosser, author, "Fast Food Nation:
The Dark Side of the All-American Meal"

Welcome to the end of the book. Although not a long journey, we have covered quite a bit of material. Short in length, but dense in content. In this last section, I thought it would be beneficial to capture the most pertinent information in the book.

First and foremost, fast food is no different than the highly processed or ultra-processed foods sold in grocery stores, big box stores, and convenience stores. The same industrialized ingredients are used. The U.S. Food & Drug Administration (FDA) has approved over 9,000 additives and the list grows every year. Of course, not all these additives are in use ... some have gone out of fashion or have been replaced. Unfortunately, I am not aware of a single on-line U.S. database that lists all the additives and their functions. Using food additives, the food industry and its supporting companies (food ingredient developers) have created foods that never existed before focusing on exemplary texture, taste, smell, eye appeal, and shelf stability. The food additives, products of food science, are miraculous substances in terms of how they can alter natural foods to make them incredibly addictive to the naïve public. But are they good for us?

New food additives are submitted to the FDA for approval, usually under the submission process called GRAS (Generally Recognized as Safe), a faster, less involved process. The onus is placed on the manufacturer or their representative to show that an additive is safe for human consumption. The FDA rarely conducts its own investigations due to limitations in staff and funding. The FDA reviews the laboratory data submitted by the food company, and, if everything is in order, grants GRAS status to the new additive. From that point, the additive may begin showing up in food ingredient labels on commercial packages. The ingredient is deemed safe until a health issue arises after commercialization. Then the FDA will investigate the problem and rescind GRAS status if the ingredient turns out to be a health hazard.

Fast food is pervasive and ever growing in our society. About a third of the U.S. population consumes fast food every day, and 11.3% of calories consumed comes from fast food. That stat is very believable if some evening you just observe the steady traffic through the drive-through lanes at a local fast-food restaurant (Chick-fil-A I think beats them all!) In 2014, there were 228,677 fast food restaurants in the U.S. In this book, I chose to examine the menus of three of the top chains: McDonald's, Taco Bell, and Pizza Hut. Fortunately for me, these companies post the ingredients in their foods and beverages at their websites. Sadly, the majority of fast-food restaurants keep that information proprietary. Although nutritional data must be provided to consumers, there is no federal law requiring restaurants to reveal ingredients. To evaluate menu items using the same basis, the author devised a simple Processed Food Index (PFI), a relative scale from 0 to 100% that measures the degree of industrialization or the "unnaturalness" of a processed food. A menu item in a fast-food restaurant with a large number of ingredients is the first clue that you've encountered a highly processed food, e.g., McDonald's Southwest Buttermilk Crispy Chicken Salad with 105 components.

Highly industrialized foods are characterized by a (1) high percentage of unique ingredients; (2) high percentage of highly processed ingredients; (3) high processed food index (PFI), typically over 40; and (4) large number of synthetic ingredients. Food manufacturers want their products to last as long as possible on store shelves ... that's called shelf life. Many food additives are designed to promote long shelf lives ... months if not years.

The FDA allows the use of a handful of acronyms to represent ingredients. Unfortunately, acronyms are not very useful to consumers, who likely won't know what they stand for. If you saw TBHQ on a label, would you know what that represents or even know why it's in the food? There are hundreds, if not thousands, of natural flavor ingredients. Companies protect them as trade secrets (proprietary formulas), and the FDA allows them to hide their identities on food labels. Many of the ingredients used to make fast food are chemically synthesized and only remotely, if at all, resemble natural ingredients that they may have been derived from. Many fast-food menu items are complex, industrially sourced mixtures that can be characterized by a Processed Food Index (PFI) to reveal the degree of industrialization. Since restaurants are not required to provide ingredients and most Americans don't know the purpose of additives, consumers are ignorant about how the additives may be undermining their health in both the short and long terms. Maybe for the first time in human history, people, particularly those in developed countries, don't know what they're eating, and quite frankly, most don't care as long as the food looks appetizing, smells great, and tastes fantastic. This book is a call out to the American public to wake up and smell the additives.

In the last 10 years, various research studies have been conducted around the world to determine whether there's an association between diets high in the consumption of highly processed and ultra-processed foods with the development of chronic diseases and mortality. So far, the studies are pretty convincing that there is a strong link. However, since the research involves prospective studies, only associations or correlations are possible and not provable causes. Given the large number of people surveyed (cohort) and the length of the studies (many years), the ever-growing number of studies point to a logical conclusion. With the expansive consumption of fast food in America and around the world, the rates of chronic diseases will continue to grow and adversely affect mortality rates. For 2015 to 2016, the Centers for Disease Control and Prevention (CDC) estimated the obesity rate for adults at 40% in the USA. Multiple studies have shown that rising rates of obesity is tied to the promotion and increased production of ultra-processed foods. Shockingly, 58% of the calories consumed in the U.S. come from ultra-processed foods.

Because of the strong associations between high consumptions of ultra-processed foods and sickness, many countries have crafted laws advocating for the placement of front-of-package labels to assist consumers to make wise choices while cruising the supermarket. Most laws are voluntary while few are mandatory. The best food scoring system for front-of-package labeling is the Nutri-Score, developed in France, which is quickly being adopted by other European countries. It addresses both the nutrient content and the degree of processing of commercial foods. Sadly, the United States has no plans to implement such a system. Even the Dietary Guidelines for Americans don't even mention ultra-processed foods or fast food. The U.S. government is incredibly negligent in addressing front-of-package labeling to help protect its citizens from making bad food choices.

So, what's the solution? First, acknowledge that there's a problem. Second, implement sound policies to reverse negative trends. To help foster public health, countries need to implement standard and mandatory systems of food package labeling that provide quick and understandable information about food quality, including both nutritional information and the degree of processing. Also, food companies, including fast food restaurants, should be required to reveal the ingredients in their products. With rising obesity rates and incidences of chronic diseases, the healthcare system is in crisis. Given the incredible advances in medical and nutritional science, it's an absolute shame that people are slowly killing themselves with their forks. The powerful influence of junk food and fast-food companies, through the production of addictive, industrial foods and the persistent and relentless marketing of their products, particularly to children, needs to be curtailed and regulated. There should be a ban on advertising of dangerous food products similar to the controls imposed on the tobacco and alcohol industries. All commercial foods should have PFIs on their labels alongside the nutritional data. More stringent laws could require food taxes on unhealthy foods to help pay for future healthcare expenses.

Without governmental intervention to improve public health, this trend is likely to continue until crisis levels of healthcare are reached.

***

From the Food Labels Revealed podcast, remember this:

"Eat foods mainly from natural plants, not manufacturing plants."

APPENDIX A:

CALCULATING THE PROCESSED FOOD INDEX (PFI)

To illustrate the calculation of a Processed Food Index (PFI), consider the example of the **Cheesy Rollup** from Taco Bell. Here are the ingredients:

Flour Tortilla: Enriched Wheat Flour(1), Water(2), Vegetable Shortening (Soybean(3), Hydrogenated Soybean(4) and/or Cottonseed Oil(5)), Sugar(6), Salt(7), Leavening (Baking Soda(8), Sodium Acid Pyrophosphate(9)), Molasses(10), Dough Conditioner (Fumaric Acid(11), Distilled Monoglycerides(12), Enzymes(13), Vital Wheat Gluten(14), Cellulose Gum(15), Wheat Starch(16), Calcium Carbonate(17)), Calcium Propionate(18), Sorbic Acid(19), and/or Potassium Sorbate(20).

Three Cheese Blend: Low-Moisture Part-Skim Mozzarella Cheese, Cheddar Cheese, Pasteurized Process Monterey Jack And American Cheese With Peppers (Cultured Milk(21), Cultured Part-Skim Milk(22), Water(23), Cream(24), Salt(25), Sodium Citrate(26), Jalapeno Peppers(27), Sodium Phosphate(28), Lactic Acid(29), Sorbic Acid(30), Color Added: Annatto(31) and Paprika Extract Blend(32), Enzymes(33), Anticaking Agents (Potato Starch(34), Cornstarch(35), Powdered Cellulose(36))).

There are 36 ingredients in this menu item. First, determine the number of unique ingredients, i.e., ones that are not repeated. Two of the ingredients, water and salt, appear twice, so there are 34 unique ingredients.

Next, identify those ingredients which are essentially unprocessed or minimally processed. These ingredients are very close to their natural state, but may undergo some simple processing treatments like washing, drying, grinding, etc. In the Cheesy Rollup, there are 3 ingredients in this category: water, salt, and jalapeno peppers.

Next, determine which ingredients are moderately, highly, or extremely processed. There are no handbooks or government tables that provide these descriptors. The author made determinations based on his knowledge and personal experience regarding the manufacturing of ingredients. However, any ingredients which are synthetic (man-made) are automatically described as extremely processed. Here is the table of determinations:

Ingredient ID (List Number)	How Processed?
Annatto (31)	Moderately
Baking Soda (synthetic) (8)	Extremely
Calcium Carbonate (synthetic) (17)	Extremely
Calcium Propionate (synthetic) (18)	Extremely
Cellulose Gum (synthetic) (15)	Extremely
Corn Starch (35)	Highly
Cottonseed Oil (5)	Highly
Distilled Monoglycerides (synthetic) (12)	Extremely
Enzymes (13,33)	Highly
Fumaric Acid (synthetic) (11)	Extremely
Hydrogenated Soybean Oil (4)	Extremely
Lactic Acid (synthetic) (29)	Extremely
Paprika Extract (32)	Moderately
Potassium Sorbate (synthetic) (20)	Extremely
Potato Starch (34)	Moderately
Powdered Cellulose (36)	Highly
Sodium Acid Pyrophosphate (synthetic) (9)	Extremely
Sodium Citrate (synthetic) (26)	Extremely
Sodium Phosphate (synthetic) (28)	Extremely
Sorbic Acid (synthetic) (19,30)	Extremely
Soybean Oil (3)	Highly
Sugar (6)	Moderately
Vital Wheat Gluten (14)	Moderately
Wheat Flour (1)	Moderately
Wheat Starch (16)	Highly

Next, count the number of ingredients in each category:

# of Moderately Processed Ingredients = 6

# of Highly Processed Ingredients = 6

# of Extremely Processed Ingredients = 13

To calculate the number of lightly processed ingredients, subtract from the number of unique ingredients the number of unprocessed, moderately processed, highly processed, and extremely processed. The unprocessed ingredients in this dish are water, milk, and jalapeno pepper.

# of Lightly Processed Ingredients = 34 – 3 – 6 – 6 – 13 = 6

Next, calculate each category as a percentage of the unique ingredients:

Unprocessed = (3/34) x 100% = 8.82%

Lightly = (6/34) x 100% = 17.7%

Moderately = (6/34) x 100% = 17.7%

Highly = (6/34) x 100% = 17.7

Extremely = (13/34) x 100% = 38.2%

Each category was weighted in terms of its contribution to the overall degree of processing. The more processed an ingredient, the greater its weight. Here are the weighting factors.

Processing Category	Weighting Factor
Unprocessed	0.000
Lightly Processed	0.010
Moderately Processed	0.100
Highly Processed	0.300
Extremely Processed	0.600

Calculate the Degree of Processing (DP) Score for each ingredient category by multiplying the percentage of unique ingredients times the weighting factor:

Unprocessed = 8.82 x 0.000 = 0.000

Lightly = 17.7 x 0.01 = 0.177

Moderately = 17.7 x 0.100 = 1.77

Highly = 17.7 x 0.300 = 5.31

Extremely = 38.2 x 0.600 = 22.9

Next, sum up the DP scores:

DP Sum = 0.000 + 0.177 + 1.77 + 5.31 + 22.9 = 30.2

The Processed Food Index (PFI) is calculated by dividing the DP Sum by 60 and expressing the result as a percentage:

PFI = (30.2/60) x 100% = 50.3 or 50.3%

NOTE: The divisor, 60, in the above equation is used since that is the highest possible DP score given that all the ingredients were extremely processed. In that example, the PFI score would be 100%. On the opposite end of the scale, if all the ingredients were unprocessed, the PFI would be 0%.

Finally, descriptors were assigned to PFI ranges to indicate to what degree the food item was industrialized. The ranges were determined by the author strictly based on reasonableness and not on a formal mathematical model.

Non-Industrialized:	0
Lightly Industrialized:	1 to 10
Moderately Industrialized:	11 to 40
Highly Industrialized:	41 to 70
Extremely Industrialized:	71 to 100

In the example of the Cheesy Rollup, the PFI value of 50.3 placing this food item in the Highly Industrialized category.

APPENDIX B:

TABLE OF MCDONALD'S FAST FOOD WITH KEY DATA

PFI = Processed Food Index

Note: This data is based on the McDonald's menu as it existed in 2020 – 2021.

ITEM	CAL. CT.	# OF INGRED.	# UNIQUE INGRED.	PFI	HOW INDUSTRIALED?
1% Low Fat Milk Jug	100	3	3	67	Highly
Apple Slices	15	2	2	50	Highly
Artisan Grilled Chicken Sandwich	430	57	44	34	Moderately
Bacon Ranch Grilled Chicken Salad	320	50	39	25	Moderately
Bacon Ranch Salad with Buttermilk Crispy Chicken	470	68	53	35	Moderately
Bacon, Egg & Cheese Bagel	550	81	63	51	Highly
Bacon, Egg & Cheese Biscuit	460	53	41	46	Highly
Bacon, Egg & Cheese McGriddles	420	47	39	48	Highly
Big Breakfast® with Hotcakes	1340	83	52	48	Highly

ITEM	CAL. CT.	# OF INGRED.	# UNIQUE INGRED.	PFI	HOW INDUSTRIALED?
Big Breakfast®	750	51	45	45	Highly
Big Mac®	540	68	52	41	Highly
Buttermilk Crispy Chicken Sandwich	600	75	54	43	Highly
Cheeseburger	300	57	41	44	Highly
Chicken McNuggets®	170	30	29	56	Highly
Chocolate Shake, Medium	630	25	19	34	Moderately
Coca-Cola®, Medium	220	6	6	39	Moderately
Creamy Ranch Dressing	110	20	20	30	Moderately
Dasani® Water	0	4	4	51	Highly
Diet Coke®, Medium	0	11	11	66	Highly
Donut Sticks, 12-Piece	730	43	34	49	Highly
Double Cheeseburger	440	50	39	42	Highly
Double Quarter Pounder® with Cheese	720	67	52	43	Highly
Dr. Pepper®, Medium	200	11	11	58	Highly
Egg McMuffin®	300	49	40	54	Highly
Fanta® Orange	210	11	11	76	Extremely
Filet-O-Fish®	380	68	51	44	Highly
Fruit & Maple Oatmeal	320	23	23	39	Moderately
Fruit 'N Yogurt Parfait	210	26	23	36	Moderately
Hamburger	250	46	36	43	Highly

ITEM	CAL. CT.	# OF INGRED.	# UNIQUE INGRED.	PFI	HOW INDUSTRIALED?
Hash Browns	140	11	11	48	Highly
Honest Kids® Appley Ever After® Organic Juice Drink	35	6	6	37	Moderately
Honey Mustard Sauce	60	20	20	37	Moderately
Hot Mustard Sauce	50	17	17	40	Moderately
Hot Cakes and Sausage	780	40	31	55	Highly
Hot Cakes	590	32	28	59	Highly
Ice Tea	0	4	4	0	Non-Industrialized
McCafe® Americano, Medium	0	1	1	2	Lightly
McCafe® Baked Apple Pie	240	24	23	49	Highly
McCafe® Cappuccino, Medium	160	4	4	26	Moderately
McCafe® Caramel Cappuccino, Medium	260	15	12	42	Highly
McCafe® Caramel Frappe´, Medium	510	36	22	42	Highly
McCafe® Caramel Latte, Medium	320	15	12	42	Highly
McCafe® Caramel Macchiato, Medium	320	28	14	44	Highly
McCafe® Caramel Mocha, Medium	380	38	25	50	Highly
McCafe® Chocolate Chip Cookie	170	33	28	47	Highly
McCafe® French	230	10	9	43	Highly

ITEM	CAL. CT.	# OF INGRED.	# UNIQUE INGRED.	PFI	HOW INDUSTRIALED?
Vanilla Cappuccino, Medium					
McCafe® French Vanilla Latte, Medium	290	10	9	43	Highly
McCafe® Hot Chocolate	450	33	25	50	Highly
McCafe® Iced Caramel Latte, Medium	270	16	13	39	Moderately
McCafe® Iced Caramel Macchiato, Medium	310	29	15	41	Highly
McCafe® Iced Caramel Mocha, Medium	330	39	26	48	Highly
McCafe® Iced Coffee	180	13	13	49	Highly
McCafe® Iced French Vanilla Coffee, Medium	170	15	15	47	Highly
McCafe® Iced French Vanilla Latte, Medium	240	11	10	39	Moderately
McCafe® Iced Latte, Medium	120	5	5	21	Moderately
McCafe® Iced Mocha, Medium	330	36	27	46	Highly
McCafe® Latte, Medium	190	4	4	26	Moderately
McCafe® Mango Pineapple Smoothie, Medium	250	26	26	35	Moderately
McCafe® Mocha	510	41	30	53	Highly

ITEM	CAL. CT.	# OF INGRED.	# UNIQUE INGRED.	PFI	HOW INDUSTRIALED?
Frappé					
McCafe® Mocha Latte, Medium	380	35	26	48	Highly
McCafe® Premium Roast Coffee, Medium	0	1	1	2	Lightly
McCafe® Strawberry Banana Smoothie, Medium	240	26	25	36	Moderately
McChicken®	400	53	39	50	Highly
McDouble®	390	57	44	41	Highly
Minute Maid® Blue Raspberry Slushie, Medium	240	14	11	48	Highly
Minute Maid® Fruit Punch Slushie, Medium	210	22	19	55	Highly
Minute Maid® Pink Lemonade Slushie, Medium	250	13	10	52	Highly
Minute Maid® Premium Orange Juice, Medium	200	2	2	0.8	Lightly
Mix by Sprite® Tropic Berry, Medium	190	7	7	62	Highly
Quarter Pounder® with Cheese Bacon	620	75	56	47	Highly
Quarter Pounder® with Cheese Deluxe	620	77	56	41	Highly
Quarter Pounder® with Cheese	510	67	52	43	Highly
Reduced Sugar Low Fat Chocolate	130	11	11	19	Moderately

ITEM	CAL. CT.	# OF INGRED.	# UNIQUE INGRED.	PFI	HOW INDUSTRIALED?
Milk Jug					
Sausage Biscuit with Egg	530	43	34	46	Highly
Sausage Biscuit	460	37	30	49	Highly
Sausage Burrito	310	54	45	56	Highly
Sausage McGriddles®	430	31	27	45	Highly
Sausage McMuffin® with Egg	480	48	39	48	Highly
Sausage McMuffin®	400	46	37	50	Highly
Sausage, Egg & Cheese McGriddles®	550	47	37	42	Highly
Side Salad	15	11	11	0	Non-Industrialized
Southwest Buttermilk Crispy Chicken Salad	500	105	74	34	Moderately
Southwest Grilled Chicken Salad	340	77	64	31	Moderately
Spicy Buffalo Sauce	30	13	11	36	Moderately
Sprite®, Medium	170	6	6	56	Highly
Strawberry Shake, Medium	590	29	23	41	Highly
Sweet 'N Sour Sauce	50	25	24	39	Moderately
Sweet Tea	110	6	5	20	Moderately
Tangy Barbeque Sauce	45	25	23	44	Highly
Vanilla Shake, Medium	590	25	18	41	Highly
World Famous Fries®, Medium	320	9	9	56	Highly

APPENDIX C:

TABLE OF PIZZA HUT'S FAST FOOD WITH KEY DATA

PFI = Processed Food Index

Note: This data is based on the Pizza Hut menu as it existed in 2020 – 2021.

PFI = Processed Food Index

PPP® = Personal Pan Pizza®

HTP = Hand Tossed Pizza

OPP™ = Original Pan Pizza™

OSC® = Original Stuffed Crust®

TNC® = Thin 'N Crispy

ITEM	CAL. CT.	# OF INGRED.	# UNIQUE INGRED.	PFI	HOW INDUSTRIALED?
American Chicken Club – PPP® Slice	180	61	48	41	Highly
Backyard BBQ Chicken – HTP Slice	140	103	62	46	Highly
Backyard BBQ Chicken – OPP™ Slice	160	100	59	46	Highly
Backyard BBQ Chicken – OSC® Slice	340	109	61	48	Highly

ITEM	CAL. CT.	# OF INGRED.	# UNIQUE INGRED.	PFI	HOW INDUSTRIALED?
Backyard BBQ Chicken – PPP® Slice	180	96	57	45	Highly
Backyard BBQ Chicken – Rectangle Slice	260	98	58	45	Highly
Backyard BBQ Chicken – TNC® Slice	130	93	56	44	Highly
Bacon Cheeseburger – PPP® Slice	200	51	37	51	Highly
Badger Special – PPP Slice	170	49	36	41	Highly
BBQ Beef – HTP Slice	140	82	49	55	Highly
BBQ Beef – PPP® Slice	160	56	41	46	Highly
BBQ Chicken – PPP® Slice	180	88	56	44	Highly
BBQ Chicken with Cheddar – PPP® Slice	150	73	58	46	Highly
BBQ Lover's – PPP® Slice	200	97	55	48	Highly
Beyond Italian Sausage – TNC® Slice	280	57	46	41	Highly
Beyond Italian Sausage – HTP Slice	300	65	50	45	Highly
Beyond Italian Sausage – OPP™ Slice	380	64	49	43	Highly
Beyond Italian Sausage – OSC®	340	73	50	46	Highly

ITEM	CAL. CT.	# OF INGRED.	# UNIQUE INGRED.	PFI	HOW INDUSTRIALED?
Slice					
BLT – PPP® Slice	170	44	34	49	Highly
BLT Salad	290	47	43	48	Highly
Blue Cheese Dipping Sauce	250	26	25	41	Highly
Breadsticks	140	17	16	62	Highly
Buffalo Burnin' Hot Bone-Out Wing	90	56	35	54	Highly
Buffalo Chicken – HTP Slice	130	79	54	46	Highly
Buffalo Chicken – OPP™ Slice	140	78	52	46	Highly
Buffalo Chicken – OSC® Slice	310	87	54	45	Highly
Buffalo Chicken – PPP® Slice	160	74	50	43	Highly
Buffalo Chicken – Rectangle Slice	240	76	51	45	Highly
Buffalo Chicken – TNC® Slice	110	71	49	44	Highly
Cheese & Mushroom – 18" Slice	360	39	32	48	Highly
Cheese 18" Slice	410	38	31	49	Highly
Cheese – Big New Yorker Slice	350	35	28	51	Highly
Cheese Dipping Sauce	250	18	18	58	Highly
Cheese – Udi's® Gluten-Free Crust Slice	110	33	28	39	Moderately
Cheese – HTP Slice	130	59	41	55	Highly

ITEM	CAL. CT.	# OF INGRED.	# UNIQUE INGRED.	PFI	HOW INDUSTRIALED?
Cheese – OPP™ Slice	140	37	30	48	Highly
Cheese – OSC Slice	310	67	42	56	Highly
Cheese – PPP® Slice	150	33	28	47	Highly
Cheese – Rectangle Slice	240	35	29	46	Highly
Cheese Sticks	150	47	40	51	Highly
Cheese – TNC® Slice	110	30	26	46	Highly
Chicken Caesar Salad	410	63	53	43	Highly
Chicken Garden Salad	420	61	57	40	Highly
Chicken-Bacon Parmesan – HTP Slice	150	109	66	44	Highly
Chicken-Bacon Parmesan – OPP™ Slice	160	108	65	45	Highly
Chicken-Bacon Parmesan – OSC® Slice	350	117	61	50	Highly
Chicken-Bacon Parmesan – PPP® Slice	180	104	62	44	Highly
Chicken-Bacon Parmesan – Rectangle Slice	270	106	63	43	Highly
Chicken-Bacon Parmesan – TNC® Slice	130	101	61	43	Highly
Cinnabon® Mini Rolls	80	69	53	55	Highly
Cinnabon® Stick	80	33	28	69	Highly

ITEM	CAL. CT.	# OF INGRED.	# UNIQUE INGRED.	PFI	HOW INDUSTRIALED?
Classic Caesar Side Salad with Chicken	310	64	53	43	Highly
Classic Caesar Side Salad	100	37	35	48	Highly
Fiesta Taco Beef – PPP® Slice	200	72	49	45	Highly
Fiesta Taco Chicken – PPP® Slice	180	75	51	44	Highly
Four Pepper Pepperoni – 18" Slice	390	62	51	50	Highly
Garden Chicken – Big New Yorker Slice	120	49	39	40	Highly
Garden Chicken – PPP Slice	160	63	48	47	Highly
Garden Fresh – Big New Yorker Slice	110	35	31	41	Highly
Garden Fresh – PPP Slice	150	49	40	49	Highly
Garden Parmesan Traditional Bone-In Slice	140	54	43	46	Highly
Hawaiian Chicken – 18" Slice	390	69	51	45	Highly
Hawaiian Chicken – Big New York Slice	350	69	51	45	Highly
Hawaiian Chicken – HTP Slice	130	69	51	45	Highly
Hawaiian Chicken – OPP™ Slice	140	68	50	44	Highly
Hawaiian Chicken – OSC® Slice	310	77	52	46	Highly

ITEM	CAL. CT.	# OF INGRED.	# UNIQUE INGRED.	PFI	HOW INDUSTRIALED?
Hawaiian Chicken – PPP® Slice	150	64	48	44	Highly
Hawaiian Chicken – Rectangle Slice	230	66	49	44	Highly
Hawaiian Chicken – TNC® Slice	110	61	47	42	Highly
Hawaiian Luau – PPP® Slice	150	57	40	50	Highly
Hawaiian – PPP® Slice	140	48	38	50	Highly
Hawaiian Teriyaki Bone-Out Wing	90	58	33	51	Highly
Hershey's® Triple Chocolate Brownie	380	37	30	46	Highly
Honey BBQ Bone-Out Wing	100	69	47	50	Highly
Honey BBQ Traditional Bone-In Wing	110	45	33	52	Highly
Italian Classic – PPP® Slice	190	54	37	48	Highly
Italian Sausage – 18" Slice	440	49	38	44	Highly
Lemon Pepper Rub Bone-Out Wing	80	58	41	46	Highly
Meat Lovers® - Big New York Slice	460	92	52	49	Highly
Meat Lovers® - Udi's® Gluten-Free Crust Slice	160	86	46	43	Highly
Meat Lovers® - HTP Slice	180	92	52	49	Highly
Meat Lovers® - OPP™ Slice	190	91	51	48	Highly
Meat Lovers® -	430	112	59	55	Highly

ITEM	CAL. CT.	# OF INGRED.	# UNIQUE INGRED.	PFI	HOW INDUSTRIALED?
OSC® Slice					
Meat Lovers® - PPP® Slice	210	87	49	48	Highly
Meat Lovers® - Rectangle Slice	310	89	50	47	Highly
Meat Lovers® - TNC Slice	150	84	48	47	Highly
Naked Traditional Bone-In Wing	80	21	19	57	Highly
Nashville Hot Bone-Out Wing	90	56	33	54	Highly
Ohio's Best – PPP® Slice	180	62	45	41	Highly
Pepperoni – 18" Slice	390	49	40	50	Highly
Pepperoni – Big NY Slice	370	45	36	51	Highly
Pepperoni – Udi's® Gluten-Free Crust Slice	120	43	33	44	Highly
Pepperoni – HTP Slice	130	69	48	54	Highly
Pepperoni Lover's® - 18" Slice	470	49	40	50	Highly
Pepperoni Lover's® - Big New York Slice	440	49	40	50	Highly
Pepperoni Lover's® - Udi's® Gluten-Free Crust Slice	150	43	33	43	Highly
Pepperoni Lover's® - HTP Slice	160	49	40	50	Highly
Pepperoni Lover's® - OPP Slice	170	48	39	48	Highly
Pepperoni Lover's®	390	57	41	51	Highly

ITEM	CAL. CT.	# OF INGRED.	# UNIQUE INGRED.	PFI	HOW INDUSTRIALED?
- OSC Slice					
Pepperoni Lover's® - PPP® Slice	180	44	37	48	Highly
Pepperoni Lover's® - Rectangle Slice	290	46	38	47	Highly
Pepperoni Lover's® - TNC® Slice	140	41	35	47	Highly
Pepperoni - OPP™ Slice	140	48	39	48	Highly
Pepperoni – OSC® Slice	340	69	49	55	Highly
Pepperoni – PPP® Slice	150	44	37	48	Highly
Pepperoni – Rectangle Slice	250	46	38	47	Highly
Pepperoni - TNC® Slice	110	41	35	47	Highly
Ranch Dipping Sauce	210	30	29	51	Highly
Ranch Rub Bone-Out Wing	80	57	35	55	Highly
Smoky Garlic Bone-Out Wing	110	89	56	50	Highly
Spicy Garlic Bone-Out Wing	110	78	47	55	Highly
Stuffed Garlic Knot	80	57	39	53	Highly
Super Supreme – PPP® Slice	210	86	50	48	Highly
Supreme – 18" Slice	430	68	51	42	Highly
Supreme – Big New York Slice	390	68	48	45	Highly
Supreme – Udi's® Gluten-Free Crust Slice	130	62	42	38	Highly

ITEM	CAL. CT.	# OF INGRED.	# UNIQUE INGRED.	PFI	HOW INDUSTRIALED?
Supreme – HTP Slice	150	68	47	45	Highly
Supreme – OPP™ Slice	160	67	46	44	Highly
Supreme – OSC® Slice	370	88	55	52	Highly
Supreme – PPP® Slice	170	63	45	44	Highly
Supreme – Rectangle Slice	270	65	45	43	Highly
Supreme – TNC® Slice	120	60	44	41	Highly
Taco Pizza – PPP® Slice	170	61	47	44	Highly
The Great Beyond – HTP Slice	310	78	57	50	Highly
The Great Beyond – OPP™ Slice	390	77	56	49	Highly
The Great Beyond – OSC® Slice	350	86	58	51	Highly
The Great Beyond – TNC® Slice	290	70	53	48	Highly
Triple Meat Italiano – 18" Slice	430	67	45	52	Highly
Tuscany® Creamy Chicken Alfredo Pasta	990	82	50	52	Highly
Tuscany® Meaty Marinara Pasta	880	67	50	42	Highly
Ultimate Cheese Lover's – HTP Slice	150	83	46	55	Highly
Ultimate Cheese Lover's – OPP™ Slice	160	83	46	53	Highly

ITEM	CAL. CT.	# OF INGRED.	# UNIQUE INGRED.	PFI	HOW INDUSTRIALED?
Ultimate Cheese Lover's – OSC® Slice	340	92	47	56	Highly
Ultimate Cheese Lover's – PPP® Slice	260	79	43	54	Highly
Ultimate Cheese Lover's – Rectangle Slice	260	81	44	53	Highly
Ultimate Cheese Lover's – TNC® Slice	130	76	41	54	Highly
Veggie Lover's® - 18" Slice	390	46	37	45	Highly
Veggie Lover's® - Big New York Slice	330	46	37	45	Highly
Veggie Lover's® - Udi's® Gluten-Free Slice	100	40	30	38	Highly
Veggie Lover's® – HTP Slice	120	67	48	50	Highly
Veggie Lover's® - OPP™ Slice	130	45	36	44	Highly
Veggie Lover's® - OSC® Slice	340	67	46	54	Highly
Veggie Lover's® - PPP® Slice	140	41	34	43	Highly
Veggie Lover's® Rectangle Slice	220	43	35	42	Highly
Veggie Lover's® - TNC® Slice	100	38	31	42	Highly
Wingstreet® Buffalo Chicken Salad	690	98	70	48	Highly
Wingstreet® Crispy Chicken Caesar	830	114	73	52	Highly

ITEM	CAL. CT.	# OF INGRED.	# UNIQUE INGRED.	PFI	HOW INDUSTRIALED?
Salad					
Wingstreet® Fries with Ketchup	500	31	30	67	Highly
Wingstreet® Straight-Cut Fries Cajun Style	510	38	36	57	Highly
Wingstreet® Straight-Cut Fries Lemon Pepper Style	510	45	39	57	Highly
Zesty Italian Salad	360	73	63	50	Highly

APPENDIX D

TABLE OF TACO BELL'S FAST FOOD WITH KEY DATA

PFI = Processed Food Index

Note: This data is based on the Taco Bell menu as it existed in 2020 – 2021.

ITEM	CAL. CT.	# OF INGRED.	# UNIQUE INGRED.	PFI	HOW INDUSTRIALED?
Baha Coolada	210	31	28	61	Highly
Beach Berry Freeze™	170	17	17	72	Extremely
Bean Burrito	350	52	36	54	Highly
Beef Burrito	430	115	76	44	Highly
Beef Quesarito	650	124	74	44	Highly
Beefy 5-Layer Burrito	490	94	64	49	Highly
Beefy Potato-Rito	420	102	71	47	Highly
Black Bean Chalupa	330	67	55	39	Moderately
Black Bean Crunch Wrap Supreme	510	78	57	49	Highly
Black Bean Quesalupa	590	97	69	47	Highly
Black Bean Quesarito	630	118	78	44	Highly
Black Beans & Rice	170	39	30	25	Moderately

ITEM	CAL. CT.	# OF INGRED.	# UNIQUE INGRED.	PFI	HOW INDUSTRIALED?
Blue Raspberry Freeze™	120	12	12	69	Highly
Breakfast Crunchwrap - Bacon	670	81	59	73	Highly
Breakfast Crunchwrap – Sausage	720	80	58	49	Highly
Breakfast Crunchwrap – Steak	660	111	55	57	Highly
Burrito Supreme® - Beef	390	112	71	48	Highly
Burrito Supreme® - Chicken	370	85	58	54	Highly
Burrito Supreme® - Steak	370	107	61	49	Highly
Chalupa Supreme® – Beef	350	73	58	41	Highly
Chalupa Supreme® – Chicken	330	65	51	47	Highly
Chalupa Supreme® – Steak	330	87	55	44	Highly
Cheesy Bean & Rice Burrito	420	103	73	43	Highly
Cheesy Fiesta Potatoes	230	51	42	43	Highly
Cheesy Gordida Crunch	500	86	66	39	Moderately
Cheesy Rollup	180	36	34	50	Highly
Chicken Chipotle Melt	190	62	48	54	Highly
Chili Cheese Burrito	370	61	49	52	Highly
Chips and Nacho	220	29	25	57	Highly

ITEM	CAL. CT.	# OF INGRED.	# UNIQUE INGRED.	PFI	HOW INDUSTRIALED?
Cheese Sauce					
Cinnabon® Delights™	160	26	23	32	Moderately
Cinnamon Twists	170	11	11	20	Moderately
Crunchwrap Supreme®	530	85	63	49	Highly
Crunchwrap Taco Supreme®	190	50	42	39	Moderately
Crunchy Taco	170	34	32	35	Moderately
Diablo Sauce Packet	0	18	18	46	Highly
Fire Sauce Packet	0	16	16	28	Moderately
Ginger Mule Freeze™	170	13	13	67	Highly
Grande Nachos – Beef	1120	110	76	43	Highly
Grande Nachos – Grilled Chicken	1070	102	67	43	Highly
Grande Nachos – Steak	1080	124	74	45	Highly
Grande Nachos – Veggies	980	86	61	46	Highly
Hash Brown	160	9	8	44	Highly
Hot Sauce Packet	0	14	14	26	Moderately
Mild Sauce Packet	0	13	13	35	Moderately
Mtn Dew® Baja Blast™	150	14	14	66	Highly
Nacho Cheese Doritos® Locos Taco	170	64	50	39	Moderately
Nacho Cheese Doritos® Locos Taco Supreme	190	80	59	41	Highly
Nachos Bell	740	81	57	45	Highly

ITEM	CAL. CT.	# OF INGRED.	# UNIQUE INGRED.	PFI	HOW INDUSTRIALED?
Grande® - Beef					
Nachos Bell Grande® - Chicken	720	73	52	50	Highly
Nachos Bell Grande® - Steak	720	95	57	47	Highly
Nachos Party Pack – Beef	2050	99	71	43	Highly
Nachos Party Pack – Chicken	1960	91	63	48	Highly
Nachos Party Pack – Steak	1980	113	70	44	Highly
Nachos Party Pack – Veggie	1770	75	55	48	Highly
Naked Chicken Chalupa	470	74	55	39	Moderately
Power Menu Bowl – Chicken	470	120	71	35	Moderately
Power Menu Bowl – Steak	480	141	75	34	Moderately
Power Menu Bowl – Veggie	430	104	66	36	Moderately
Quesadilla – Cheese	470	67	55	52	Highly
Quesadilla – Chicken	510	83	61	49	Highly
Quesalupa – Beef	610	103	71	50	Highly
Quesalupa – Chicken	580	94	70	51	Highly
Quesalupa – Steak	610	117	70	51	Highly
Quesarito – Beef	650	125	76	46	Highly
Quesarito – Chicken	620	117	74	47	Highly
Quesarito – Steak	630	139	76	47	Highly
Soft Taco – Beef	180	49	43	50	Highly

ITEM	CAL. CT.	# OF INGRED.	# UNIQUE INGRED.	PFI	HOW INDUSTRIALED?
Soft Taco – Chicken	160	40	34	59	Highly
Soft Taco Supreme® – Beef	210	65	51	48	Highly
Soft Taco Supreme® – Chicken	180	56	45	53	Highly
Spicy Potato Soft Taco	230	65	52	46	Highly
Wild Strawberry Freeze™	150	20	20	55	Highly
Wild Strawberry Lemonade Freeze	180	31	24	59	Highly

APPENDIX E

FOOD PROCESSING TERMS & DEFINITIONS

TYPE OF ADDITIVE	DEFINITION
Acidity Regulator	An additive that helps to maintain the acidity level in a food.
Acidulant (Acidifier)	A substance that increases the acidity of a food.
Anti-oxidant	A substance that chemically combines with oxygen to prevent rancidity of oils.
Anti-Caking Agent	These additives keep powders free flowing and prevents particles from sticking together.
Buffer	Additives that adjust and maintain the pH (acidity or alkalinity) of a food.
Bulking Agent	Additives that increase the bulk of a food without affecting its taste.
Chelating Agent	A food additive, like EDTA, that acts as a preservative. They sequester (tie up) metals preventing them from causing color or flavor deterioration via enzymatic activity.
Defoamer (Antifoamer)	A surface-active additive that is used to control the amount of foaming during the processing of a food or beverage.
Dough Conditioner	Any ingredient added to bread dough to strengthen its texture or improve it in other ways. These additives may include enzymes, salts,

TYPE OF ADDITIVE	DEFINITION
	emulsifiers, and other chemicals.
Effervescent	Any substance that aids in forming bubbles in a food or beverage.
Emulsifier	An additive that keeps water and oil mixtures from separating.
Enrichment	A process whereby a nutrient is added back into a food to replace what was lost in processing; it is usually a synthetic chemical.
Extender	An added ingredient used to increase the bulk of a food.
Fat Substitute	A food additive, usually synthetic, with properties similar to regular fat but it contains fewer or zero calories. They are used to make low-fat and low-calorie foods.
Filler	Food fillers are additives that help bulk up the weight of a food with cheaper ingredients to keep the price of the food down.
Firming Agent	This is a chemical added to a food in order to precipitate residual pectin (gel material), thus strengthening the structure of the food and preventing its collapse during processing.
Flavor Enhancer	A chemical agent added to savory foods to boost flavor.
Gelling Agent	A food additive used to thicken and stabilize different types of foods. The agents provide the foods with texture by forming a gel, a complex network.
Humectant	An additive that helps preserve the moisture content in foods.
Leavening Agent	Any substance that serves to raise the dough in a

TYPE OF ADDITIVE	DEFINITION
	baked good.
Neutralizing Agent	An additive to shift the pH of a product from an acid or alkaline condition to neutral.
Pickling Agent	An additive that assists in the pickling of foods to preserve them.
Preservative	Inhibits mold, fungi, and yeast to increase the shelf life of a food.
Processing Aid	A substance used to improve the production of a variety of foods but which is not present in any significant amount in the finished product. It does not affect appearance or taste and, most importantly, has no impact on health.
Sequestrant	An additive that binds with offending substances in foods to prevent objectional changes (color, flavor, etc.) that could affect food quality.
Stabilizer	A substance that provides texture and consistency to a food.
Sweetener	A natural or artificial substance that enhances the sweetness of a food.
Texturizer	A chemical to improve the texture of various foods, for example, calcium chloride keeps tomatoes firm.
Thickener	A gum or starch that provides body to food mixtures.

APPENDIX F

GLOSSARY OF FOOD INGREDIENTS & ADDITIVES

NOTES: Only ingredients and additives that are classified as moderately to extremely industrialized are included in this table. Unprocessed or lightly processed ingredients are not included. Additives that are designated "synthetic" are artificial and manufactured in factories; they are automatically labelled as extremely processed. The degree of processing and the synthetic nature of an additive is included with the Ingredient Name. All the substances in this table can be found in the menu items for McDonald's, Pizza Hut, and Taco Bell restaurants. Refer to Appendix E for definitions of terms, (e.g., texturizer, leavening agent, etc.), that appear in the glossary.

INGREDIENT NAME [PROCESSING INFO]	EXPLANATION
Acesulfame Potassium [Synthetic; Extremely]	A crystalline substance used as an artificial, non-caloric, high-intensity sweetener, which is 200 times sweeter than table sugar. It is commonly combined with another sweetener, e.g., aspartame, found in the product Equal. The substance is also listed on labels as Acesulfame K (the "K" is the chemical symbol for potassium). This ingredient was first approved for use in foods in 1988.
Acetic Acid [Synthetic; Extremely]	Acetic Acid is a natural substance found in fruits and other foods, but a synthetic version is generally used in processed foods. It is the sour substance in vinegar where it is present in amounts from 4 to 6%. Acetic acid is used as an acidulent, flavoring agent, preservative, and pickling

INGREDIENT NAME [PROCESSING INFO]	EXPLANATION
	agent.
Alum [Synthetic; Extremely]	Alum is composed of potassium aluminum sulfate, which is a common pickling agent. The chemical is hazardous but only in concentrated solutions. Note that aluminum is not natural to the human body and has been linked to Alzheimer's disease.
Ammonium Bicarbonate [Synthetic; Extremely]	This chemical is used as a leavening agent for making baked goods.
Ammonium Sulfate [Synthetic; Extremely]	In baked goods, this chemical is a food for yeast, a dough conditioner, and a buffer to control acidity.
Annatto [Moderately]	This is a vegetable dye (yellow to pink) and a spice flavoring (sweet, peppery) obtained from the seeds of a tropical tree. The seeds are processed to extract the color component. Annatto may cause allergic reactions in some people.
Annatto Extract [Moderately]	[See Annatto]
Apo-Carotenal [Synthetic; Extremely]	This additive, related to vitamin A, is a crystalline, orange-red powder used to color beverages, cheese, desserts, and ice cream, particularly in fat-heavy foods. Although found in vegetables, like spinach, the industrial additive is man-made.

INGREDIENT NAME [PROCESSING INFO]	EXPLANATION
Artificial Flavors [Synthetic; Extremely]	This generic term could represent thousands of substances that the Food & Drug Administration (FDA) has approved for use in foods and beverages. "Artificial" means that the additive has been synthesized in a factory. Major research efforts go into the creation of artificial flavors. Some of them are used to mimic real foods, like flavors for "roast chicken", "grilled meat", or "fruity". Also, artificial flavors may have no taste on their own but enhance flavors like sweetness or savoriness. There are some big companies in this multi-billion-dollar business sector, e.g.; International Flavors & Fragrances, Firmenich, Symrise, Innova, and Sensient, that are not well known outside the food industry.
Ascorbic Acid [Synthetic; Extremely]	Another name for ascorbic acid is Vitamin C, a natural product, but its presence in a baked good is not for nutritional purposes. This generally synthesized additive is a potent anti-oxidant and aids in preserving foods.
Ascorbyl Palmitate [Synthetic; Extremely]	This chemical is a derivative of ascorbic acid and is used as a preservative and antioxidant.
Aspartame [Synthetic; Extremely]	Aspartame is an artificial sweetener which is 200 times sweeter than table sugar. It was first synthesized in 1965 and was later approved for food and beverage use in 1981 by the Food & Drug Administration. This substance is made from two amino acids, aspartic acid and phenylalanine. Commercially, it is known by the brand names Nutrasweet and Equal. As a non-nutritive sweetener, it contributes zero calories in food preparations.
Autolyzed Yeast Extract [Highly]	This substance is a flavor enhancer. It is made by suspending yeast in salt water. The salt causes the yeast cells to shrivel up and release enzymes that break the cells down. The water-soluble part of the cells stays in solution and the solid cell parts are removed. The excess water is

INGREDIENT NAME [PROCESSING INFO]	EXPLANATION
	stripped away to form a liquid paste or the product may be spray dried to produce a powder. In the name, the word "autolyzed" means self-destroying, and is followed by the word "extract" which means the soluble portion of the yeast cells have been removed from the insoluble parts. People who are sensitive to monosodium glutamate (MSG) should avoid autolyzed yeast extract since MSG is a component of this additive.
Azodicarbonamide [Synthetic; Extremely]	A chemical that serves as a whitening agent for flour and as a dough conditioner. When heated, such as in bread making, the chemical will decompose. A breakdown product is semicarbazide, which has been linked to tumor formation in rodent trials. Because of that issue, the FDA has set a limit for the use of the chemical in the food industry.
Baking Powder [Synthetic; Extremely]	Baking powder is a mixture of leavening agents used to prepare baked goods in non-yeast products. The leavening agents can vary but a typical combination is baking soda and potassium bitartrate. The mixture produces carbon dioxide in contact with water which causes dough to rise.
Baking Soda [Synthetic; Extremely]	The chemical name is sodium bicarbonate. It is a common leavening agent used in baked goods. There are natural sources of sodium bicarbonate, e.g., the mineral nahcolite, but fast-food manufacturers are most likely going to use the cheaper, synthetic version. [See Sodium Bicarbonate]
Barley Malt Extract [Moderately]	Barley Malt Extract is derived from barley grain. It is produced by steeping barley, allowing the naturally present enzymes to convert starches into soluble sugars. The resulting syrup is then concentrated to produce an extract of malted barley. It has a unique taste and is often substituted for sugar in recipes.

INGREDIENT NAME [PROCESSING INFO]	EXPLANATION
Beta Carotene [Highly]	As a compound with a red-orange pigment, this chemical is often used as a coloring agent. It can be extracted from algae. The key component is separated using a non-polar solvent such as hexane.
Bleached Wheat Flour [Highly]	Bleaching wheat flour changes some physical and chemical properties of the flour to aid in different baking processes, but the main reason for bleaching is to whiten flour. Natural wheat flour has a light yellow or brown tint to it due to the presence of proteins. There are a variety of chemical agents available to food manufacturers for bleaching purposes. Some are hazardous.
Blue 1 [Synthetic; Extremely]	This substance is also known as Brilliant Blue. It's an artificial dye that imparts a blue color to foods and beverages. Some people may be allergic to it.
Brominated Vegetable Oil [Synthetic; Extremely]	Abbreviated as BVO, this additive is a complex mixture of plant-derived triglycerides (fats) that have been chemically reacted with elemental bromine to form brominated molecules. Its primary use is as an emulsifier in citrus-flavored soft drinks. BVO was first used by the soft-drink industry in 1931.
Brown Rice Flour [Moderately]	[See Rice Flour]
Brown Sugar [Moderately]	A modified cane sugar combined with molasses to produce a favorable color and flavor in baked goods.
Caffeine [Moderately]	Caffeine, also called trimethylxanthine, is a stimulant added to Coca-Cola and other soft drinks. It is found naturally in many plants from which it can be extracted. Also, it can by synthetically made from urea. The author is not aware of the source of the caffeine in Coca-Cola, so it is described as "moderately" processed.

INGREDIENT NAME [PROCESSING INFO]	EXPLANATION
Calcium Ascorbate [Synthetic; Extremely]	This chemical is a preservative (anti-oxidant) prepared from the combination of ascorbic acid (vitamin C) and calcium carbonate. The addition of the calcium compound diminishes the acidity of the vitamin C allowing it to be more easily absorbed and also reducing the acidity or tartness of the food.
Calcium Carbonate [Synthetic; Extremely]	Calcium Carbonate is a white powder obtained naturally from limestone or marble or synthetically by a chemical reaction. It's also found in shells. In foods, it functions as an acidity regulator, anticaking agent, and a color agent.
Calcium Chloride [Synthetic; Extremely]	This synthetic chemical is most frequently obtained by reacting limestone with hydrochloric acid. The processing of the product mixture produces a crystalline substance. It serves as a sequestrant to improve the quality and stability of foods.
Calcium Disodium EDTA [Synthetic; Extremely]	The acronym EDTA stands for "Ethylene Diamine Tetraacetic Acid." It's a preservative to prevent crystal formation and to retard color loss.
Calcium Lactate [Synthetic; Extremely]	This chemical serves as a buffer in a food product. A buffer controls the acidity or alkalinity of a product by maintaining a stable pH.
Calcium Peroxide [Synthetic; Extremely]	This is a man-made, crystalline additive used in the bakery industry as a dough conditioner, a bleaching agent for oils, and a starch modifier.
Calcium Propionate [Synthetic; Extremely]	This chemical is a preservative to keep baked goods from getting moldy. There are some downsides to the consumptions of this additive: (1) it has the potential to permanently damage your stomach lining by exacerbating gastritis and inducing severe ulcers and (2) it has been linked to migraine headaches. Children exposed to this additive may exhibit irritability, restlessness, inattention and sleep disturbances.

INGREDIENT NAME [PROCESSING INFO]	EXPLANATION
Calcium Sodium EDTA [Synthetic; Extremely]	[See Calcium Disodium EDTA]
Calcium Stearate [Synthetic; Extremely]	This chemical additive is produced by heating stearic acid and calcium oxide. It is commonly found as the main, insoluble component of soap scum, a white solid that forms when soap is mixed with hard water. In the food industry it is used as an anti-caking agent.
Calcium Sulfate [Synthetic; Extremely]	Commonly known as Plaster of Paris, it comes from a clear, white rock found in nature. It is used as a firming agent, a food for yeast, and as a dough conditioner for baked goods. Commercial bakeries employ various dough conditioners for several reasons: (1) to shorten dough rising times (2) to increase shelf life and (3) make the dough easier for machinery to process.
Canola Oil [Highly]	This material comes from the seeds of the rapeseed plant, which is related to turnips. To produce the oil, the seeds are slightly heated and then crushed. The oil is then extracted using the petroleum solvent hexane. For the finished product, the oil is refined using a water wash to remove unwanted impurities, then filtered to remove color, and lastly deodorized using steam distillation. In North America, most canola oil comes from Canada. Since the name "rapeseed" conjures up a negative image, the Canadian industry renamed the oil to "canola" which is a condensed word obtained from "**Can**ada" and "**o**il" and "**l**ow **a**cid." In the early days of its use, rapeseed oil contained high amounts of a toxic chemical called erucic acid. Current processing methods remove erucic acid.
Caramel [Highly]	[See Caramel Color]
Caramel Color [Highly]	This coloring agent has been around a long time in the food industry. It's what gives Coca Cola its characteristic

INGREDIENT NAME [PROCESSING INFO]	EXPLANATION
	deep-brown color. It's manufactured by heating sugar in the presence of acid or alkali in what is known as a caramelization process.
Carbonated Water [Moderately]	Carbon dioxide gas is dissolved in water under pressure to form carbonated water.
Carnauba Wax [Moderately]	Carnauba wax, often referred to as Brazil wax, is a product obtained from the leaves of the carnauba palm tree, a plant native to and grown only in a northeastern Brazilian state. The refined material is used as a coating for candies.
Carrageenan [Moderately]	This additive is extracted from red edible seaweeds (Irish moss). The seaweed is boiled in water, and then the carrageenan extract is precipitated with cold water. It is widely used in the food industry for its gelling, thickening, and stabilizing properties. The main application is in the dairy and meat industry due to its strong binding to food proteins.
Cellulose [Moderately]	[See Cellulose Powder]
Cellulose Gel [Synthetic; Extremely]	[See Microcrystalline Cellulose]
Cellulose Gum [Synthetic; Extremely]	This material is sometimes written on labels as sodium carboxymethylcellulose (CMC). It's a chemically modified form of cellulose, which is the carbohydrate found in woody plants, like trees, and also in cotton. Cellulose gum is used as a stabilizer in foods.

INGREDIENT NAME [PROCESSING INFO]	EXPLANATION
Cellulose Powder [Moderately]	This versatile and common food additive is obtained from the cell walls of plants (trees, vegetables, fruits, etc.). Powdered cellulose is made by cooking raw plant fiber, like wood, in various chemicals to separate the cellulose, and then it is purified. It is a source of insoluble fiber. It has a wide variety of uses such as adding creaminess, increasing firmness, retarding spoilage, serving as a binder in meat, a filler, and can be used as a coating to keep food from sticking together. This chemical may also be labeled as microcrystalline cellulose. Cellulose powder also helps to incorporate air in baked goods. The general uses are as a texturizer, an anti-caking agent, a fat substitute, an emulsifier, an extender, and a bulking agent in food production.
Cheese Whey [Moderately]	[See Whey]
Citric Acid [Synthetic; Extremely]	Although found in nature, citric acid, as a food ingredient, is generally manufactured in a chemical plant using complicated industrial processes and hazardous chemicals. This additive shows up in many food products, e.g., cheese sauces. In baked goods, it creates a slightly acidic environment for yeast to ferment more effectively, thus decreasing the time it takes for dough to rise. In meats, it assists in curing. It can help remove off-flavors in food products.
Coffee Extract [Moderately]	Coffee extract is made by brewing coffee beans in alcohol.
Color Extractive of Annatto [Highly]	[See Annatto]
Color Extractive of Turmeric [Highly]	[See Extractives of Turmeric]

INGREDIENT NAME [PROCESSING INFO]	EXPLANATION
Confectioner's Glaze [Moderately]	Glaze is an alcohol-based solution of food-grade shellac, which is obtained from a resin excreted by the lac bug, native to India and Thailand. It provides a shiny and protective coating to candies.
Corn Flour [Moderately]	This flour comes from degerminated corn that has been dried and finely ground. Degerminated corn is made by removing the nutritious, oily germ and bran during processing so that the final product is shelf stable.
Corn Maltodextrin [Highly]	[See Maltodextrin]
Corn Oil [Highly]	This material comes from the seeds of corn plants, usually the variety called dent corn. The corn oil is expeller-pressed, then extracted using hexane. After extraction, the oil is refined by degumming and by treatment with alkali to remove impurities and color. The last stages of refinement include the removal of waxes and the elimination of odor using steam distillation under high vacuum.
Corn Starch (Cornstarch) [Highly]	This material is manufactured by the breakdown of corn kernels in a process called wet milling. Corn is heated (steeped) in a slightly acidic mixture. The endosperm (starch part) of the corn is separated from the germ (oil part). The endosperm fraction is ground and the starch component is removed by washing. Further separations using centrifuges remove impurities like fiber. Finally, the white starch is dried.
Corn Syrup [Highly]	Various syrups (sweeteners) can be created as the result of breaking down cornstarch. If the food label doesn't say High Fructose Corn Syrup (HFCS), then the other corn sweeteners get lumped into this generic category. See High Fructose Corn Syrup.
Corn Syrup Solids [Synthetic;	These substances are derived from the dehydration (water removal) of corn syrup. See Corn Syrup.

INGREDIENT NAME [PROCESSING INFO]	EXPLANATION
Extremely]	
Cottonseed Oil [Highly]	Cottonseed Oil is removed from the seed of the plant using solvent extraction, a physical process. Then it is refined by the following process: (1) degumming with water and dilute acids, (2) neutralization, (3) bleaching using bentonite clay, (4) decolorization, and (5) winterization to prevent solidification. This oil can be found in shortenings, fried foods, and salad oils.
Cultured Corn Syrup Solids [Synthetic; Extremely]	[See Corn Syrup Solids] Bacterial cultures are added to corn syrup solids to produce anti-bacterial compounds in a fermentation-type process. The resultant material is dried and added to processed foods as a preservative.
Cultured Wheat Flour [Moderately]	Cultured Wheat Flour is a natural preservative made from the controlled fermentation of wheat flour using the same bacterium that Swiss cheese manufacturers use. This ingredient helps to extend the shelf-life of baked goods and other foods by fending off molds and bacteria. It can help to replace synthetic preservatives like sorbates and benzoates to make cleaner (healthier) products.
Cyanocobalamin [Synthetic; Extremely]	This additive is vitamin B12. It is essential for forming red blood cells and general neurological health.
DATEM [Synthetic; Extremely]	This acronym stands for Diacetyl Tartaric Acid Esters of Mono- and Diglycerides. To understand where this synthetic ingredient comes from would require a degree in chemistry. Functionally, it serves as an emulsifier for keeping water and oily components from separating, similar to eggs in mayonnaise. It's also a dough conditioner used to improve volume and uniformity. It is considered safe by the FDA, but a study in 2002, on rats, showed "heart muscle fibrosis and adrenal overgrowth".
Degermed Yellow Corn Flour	This is corn flour made from yellow corn in which the germ (oil component) has been removed. In the absence

INGREDIENT NAME [PROCESSING INFO]	EXPLANATION
[Moderately]	of the oily component, the corn flour will last longer without refrigeration.
Dehydrated Butter [Moderately]	Butter is dehydrated to give a powdered form.
Dextrin [Highly]	Dextrins are mixtures of substances derived from lightly hydrolyzed starch using enzymatic or acid/heat methods. They can enhance crispness in food batters.
Dextrose [Synthetic; Extremely]	This ingredient is more commonly known as glucose. Although natural, dextrose is a very industrial food ingredient typically sourced from corn. To briefly describe how it's made, cornstarch is separated from corn kernels, and then a liquid mixture of the cornstarch is broken down by acid or enzymes or a combination of the two at elevated temperatures and pressures until a liquid mixture of dextrose is formed. Then the water is removed to give crystalline dextrose, a white powder. Dextrose is a common ingredient found in many food products and its main function is to serve as a sweetener.
Diacetyl Tartaric Acid Ester of Monoglyceride [Synthetic; Extremely]	[See DATEM]
Dicalcium Phosphate [Synthetic; Extremely]	Dicalcium Phosphate is a white, tasteless, microcrystalline powder that is frequently utilized in foods and beverages as a buffering agent, a dough modifier, an emulsifier, a leavening agent, a nutritional supplement, or a stabilizer. People with chronic kidney disease may need to watch their calcium phosphate intake, however, as it could adversely affect their health if they consume too much phosphate.
Dimethylpolysiloxane [Synthetic;	This chemical, also called polydimethylsiloxane (PDMS), is derived from silicone oil. It's used as a lubricant, anti-

INGREDIENT NAME [PROCESSING INFO]	EXPLANATION
Extremely]	foaming agent, and preservative. It acts as a preservative since it is not biodegradable.
Dipotassium Phosphate [Synthetic; Extremely]	This additive is a white powder commonly used in foods as an emulsifier, stabilizer, texturizer, and sequestrant.
Disodium Guanylate [Extremely]	Disodium Guanylate is a flavor enhancer. Flavor enhancers (see also disodium inosinate below) have little taste of their own, but they accentuate the natural flavors of foods. These chemicals were first isolated in Japan in 1913 from fish. Food manufacturers often use them in combination with monosodium glutamate (MSG) because there is a synergistic effect. These additives are considered safe, but both of them can get converted to uric acid in the body, so people suffering from gout should avoid foods that have these chemicals.
Disodium Inosinate [Extremely]	Disodium Inosinate is another flavor enhancer. See Disodium Guanylate.
Disodium Phosphate [Synthetic; Extremely}	This chemical has various uses such as a sequestrant, an emulsifier, and a buffer to control acidity.
Distilled Monoglycerides [Synthetic; Extremely]	Purified monoglycerides. [See Mono- and Diglycerides]
Distilled Vinegar [Moderately]	[See Vinegar]
Dried Cane Syrup [Moderately]	[See Evaporated Cane Sugar]
Dried Vinegar [Highly]	Vinegar is normally a weak solution of acetic acid in water. So how does a liquid ingredient get turned into a dry product? Here's how it's done. A thin layer of vinegar is sprayed onto maltodextrin or a modified food starch [see Glossary listings for these materials]. The porous

INGREDIENT NAME [PROCESSING INFO]	EXPLANATION
	structure of the maltodextrin or starch absorbs a large amount of vinegar. The excess liquid is removed by a drying process to produce a powder, which can then be used in any food where a dry vinegar is needed. Vinegar is used for flavor and to control acidity in food products.
Enriched Wheat Flour [Moderately]	[See Wheat Flour]
Enzymes [Highly]	This is not a specific food additive but a class of organic proteins that serve as catalysts for biochemical reactions. The most common enzyme for baked goods is amylase, which breaks down starch into simple sugars. Amylase aids in the rising of wheat dough to produce a desirable volume. It also helps to develop a pleasing crust color, contributes to taste, and improves the shelf life of the product. This additive is another example of how the Food & Drug Administration, in some cases, allows manufacturers to list a class of additives instead of specific chemical names. It's hard to know how many enzymes are used in processed foods, but it would be at least dozens, if not hundreds.
Erythorbic Acid [Synthetic; Extremely]	Chemically related to ascorbic acid (vitamin C), this synthetic additive is used as an antioxidant and preservative in processed foods.
Ester Gum [Synthetic; Extremely]	[See Glycerol Ester of Wood Rosin]
Ethylated Mono- and Diglycerides [Synthetic; Extremely]	The McDonald's website likely misnamed this ingredient. The additive should have been listed as eth**ox**ylated mono- and diglycerides. These chemicals are emulsifiers produced by an interaction between glycerides (fat molecules) and the poisonous gas ethylene oxide. Their dough strengthening properties are strong, so they are

INGREDIENT NAME [PROCESSING INFO]	EXPLANATION
	popular for crusty breads with a chewy texture. They also improve shelf life for baked goods. These additives are used to prolong flavor freshness and increase loaf volume in breads.
Evaporated Cane Sugar [Moderately]	A less-processed type of cane sugar which retains more nutrients than regular white sugar. Used as a sweetener.
Expeller-Pressed Canola Oil [Moderately]	This is a less processed and healthier form of canola oil. [See Canola Oil]
Extractives of Black Pepper [Highly]	This is a specific example of a spice extractive. [See Spice Extractives]
Extractives of Celery [Highly]	This is a specific example of a spice extractive. [See Spice Extractives]
Extractives of Paprika [Highly]	This is a specific example of a spice extractive. [See Spice Extractives]
Extractives of Rosemary [Highly]	This is a specific example of a spice extractive. [See Spice Extractives]
Extractives of Turmeric [Highly]	This is a specific example of a spice extractive. [See Spice Extractives]
Extra-Virgin Olive Oil [Highly]	[See Olive Oil]
FD&C Blue #1 [Synthetic; Extremely]	[See Blue 1]
FD&C Red #40 [Synthetic; Extremely]	[See Red 40]

INGREDIENT NAME [PROCESSING INFO]	EXPLANATION
Fava Bean Protein [Highly]	Protein isolates from soybeans have been around a long time now, but recently other legumes are being used to produce protein isolates. These additives help with water and fat absorption, emulsification, aeration, and for increasing protein content. With the exception of soy, information on the processing of specific legumes is difficult to find due to proprietary information. However, here is a general processing outline: (1) Ground legumes are defatted to produce a flaky material; (2) The flakes are mixed with a near neutral extraction medium in agitated, heated vessels; (3) The protein extract is clarified by filtration; (4) The protein fraction is precipitated by acidifying; and (5) The protein curds are washed and then dried.
Ferrous Gluconate [Synthetic; Extremely]	An additive for fortification of white flour to provide iron, an essential mineral.
Ferrous Sulfate [Synthetic; Extremely]	An additive for fortification of white flour to provide iron, an essential mineral.
Flavor(s) [Synthetic; Extremely]	[See Artificial Flavors]
Folic Acid [Synthetic; Extremely]	This is the chemical name for Vitamin B9 and is used to enrich white bread products.
Food Starch [Highly]	Generic ingredient. See cornstarch, tapioca starch, wheat starch, etc.
Food Starch Modified [Synthetic; Extremely]	[See Modified Food Starch]
Fructose	Fructose is also called "fruit sugar" because it is present

INGREDIENT NAME [PROCESSING INFO]	EXPLANATION
[Synthetic; Extremely]	in most fruits and provides a naturally sweet taste. The common form of it is crystalline fructose obtained from a complex, industrial process where dextrose is converted to fructose and then purified and dried to produce crystals. It serves as a sweetener and is 150% sweeter than table sugar (sucrose).
Fumaric Acid [Synthetic; Extremely]	An acidulant which helps control acidity in foods and provides a sour taste. It is made from a related compound, maleic acid, by means of catalytic isomerization.
Gelatin [Extremely]	The main ingredient in Jell-O, gelatin is a translucent, colorless, flavorless food ingredient, derived from collagen (found in connective tissue) taken from slaughtered pigs and cows. It is commonly used as a gelling agent in food. The process for making gelatin is long and complex involving many steps.
Gellan Gum [Moderately]	This substance is used as a stabilizer and thickener in various foods. It is produced by a bacterium discovered in 1978 in the tissue of a lily plant in Pennsylvania, USA.
Glucono Delta-Lactone [Synthetic; Extremely]	This chemical has multiple uses in the food industry as a leavening agent, acidifier, pickling agent, and sequestrant.
Glycerin or Glycerol [Synthetic; Extremely]	This substance is a breakdown product of fats and oils. It is used as a humectant, a solvent for colors and flavors, and as a plasticizer in edible coatings.
Glycerol Ester of Rosin [Synthetic; Extremely]	Glycerol ester of rosin is also known as glycerol ester of wood rosin, glyceryl abietate, or ester gum. It is an oil-soluble food additive. When used in beverages it acts as a stabilizer to keep oils in suspension in water. Its purpose is to mix together water and flavoring oils from fruits. To make the substance, refined wood rosin, from the stumps of long-leaf pine trees, is reacted with glycerin (glycerol).

INGREDIENT NAME [PROCESSING INFO]	EXPLANATION
Grape Vinegar [Moderately]	[See Vinegar]
Green 3 [Synthetic; Extremely]	This substance is also known as Fast Green. It's an artificial dye that imparts a green color to foods and beverages. Some people may be allergic to it.
Guar Gum [Moderately]	A thickening agent in foods that is extracted from guar beans. The guar seeds are mechanically dehusked, hydrated, milled and screened according to application. It is typically produced as a free-flowing, off-white powder.
High Fructose Corn Syrup (HFCS) [Synthetic; Extremely]	HFCS is one of the most common, nutritive sweeteners used in the food industry. This industrial food ingredient is derived from cornstarch using a complex industrial process. HFCS is a great example of the application of chemistry to create a product that had never existed before in foods prior to the 1970s, plus, in a few short years, it rapidly replaced sugar in many products, particularly sweetened beverages. Is it good, bad, or neutral for you? HFCS is probably not much worse than sugar, but some people say it has some detrimental properties, like increased weight gain, that may affect the rate of obesity in this country, but that's still a pretty controversial subject.
Hydrogenated Oils (Soybean Oil, Palm Oil, Cottonseed Oil, Coconut Oil) [Synthetic; Extremely]	To prepare a hydrogenated oil, the pure oil is reacted with hydrogen gas in the presence of a metal catalyst. In the process, partially unsaturated fats are converted to saturated fats, which are more shelf stable. If the process is only partially completed, then trans-fats will form.
Hydrolyzed Corn, Soy, Wheat [Highly]	When proteins from corn, soy, or wheat are subjected to breakdown by acid at elevated temperatures and the resultant mixture is neutralized, then umami (savory) flavors are formed, e.g., non-fermented soy sauce. These

INGREDIENT NAME [PROCESSING INFO]	EXPLANATION
	materials are used as flavoring agents.
Hydrolyzed Vegetable Protein, Corn & Soy [Highly]	[See Hydrolyzed Corn, Soy, Wheat]
Hydroxy Propyl Methylcellulose [Synthetic; Extremely]	A chemical used as an emulsifier and to increase viscosity (thickness). It can serve as an alternative to gelatin.
Interesterified Soybean Oil [Synthetic; Extremely]	Interesterified Soybean Oil is a type of oil where the parts of the oil molecules are chemically rearranged to modify properties like the melting point, slow down rancidity rate, and to make an oil more stable for deep frying. Some research suggests that these new fats are just as bad for you as trans fats. They might increase heart-disease risk by lowering HDL (good) cholesterol and raising LDL (bad) cholesterol, as trans fats do. And they might increase the risk of type 2 diabetes by raising fasting blood-glucose levels and decreasing insulin response.
Invert Sugar [Synthetic; Extremely]	This is a type of sweetener made from sucrose (table sugar). Sugar dissolved in water is broken down at an elevated temperature into dextrose and fructose using either enzymes, hydrochloric acid, or cream of tartar. The resulting liquid (inverted sugar syrup) is slightly sweeter than the original sugar solution.
Iron [Synthetic; Extremely]	Iron is an elemental metal that is mandated for use in refined wheat flours for fortification. However, elemental iron is not used directly in food products, but is chemically reduced (ionized) to an alternative form, e.g., Ferrous Gluconate, Ferrous Sulfate, and Reduced Iron (generic descriptor).
Konjac Flour [Moderately]	Konjac, also known as glucomannan or voodoo lily or devil's tongue, is a plant that grows in Southeast Asia. It's known for its starchy corm, a tuber-like part of the stem

INGREDIENT NAME [PROCESSING INFO]	EXPLANATION
	that grows underground. The corm produces a source of soluble, dietary fiber. To obtain the flour, mature corm is dried and then processed into a powder. Konjac is used as a gelatin substitute and to thicken or add texture to foods.
L-Cysteine [synthetic; extremely]	L-Cysteine is an essential amino acid derived from hair. It is used as a dough conditioner in baked goods. As a food additive, it is artificially produced.
Lactic Acid [Moderately]	Lactic acid is primarily used as an acidulant. It is manufactured by bacterial fermentation of whey, cornstarch, potatoes, or molasses.
Lactose [Moderately]	Lactose is commonly called "milk sugar" since it provides sweetness in mammal's milk. It is isolated from whey liquid, a by-product of cheese making. It is used in culture media and as a stabilizer.
Liquid Sugar [Moderately]	Liquid sugar is a product that is created using a mixture of water and white or brown sugar. There are commercially produced versions available for use in the home as well as in restaurants. Many of the commercial products will use a specific balance of glucose and fructose products in order to create a viable product. This often yields a product that is actually sweeter than regular table sugar.
Locust Bean Gum (Carob Bean Gum) [Highly]	This is one example of a variety of vegetable gums that are used in food products. Gums are used to give foods body, a gel-like consistency. Essentially, they are thickeners. Locust bean gum is extracted from the seeds of the carob tree found in the Mediterranean region. It's an off-white powder.
Magnesium Sulfate [Synthetic; Extremely]	This chemical is a mineral supplement that provides magnesium.
Malic Acid	The word malic is derived from malum, Latin for apple.

INGREDIENT NAME [PROCESSING INFO]	EXPLANATION
[Synthetic; Extremely]	Malic acid is a natural product in apples which contributes a sour taste. However, the source of malic acid used in food formulas is not likely to be apples, but is more likely to come from a complex industrial chemical process. Malic acid serves as an acidulant to provide a little acidity to food products. The lower pH provided by malic acid can also help preserve food.
Malted Barley Flour [Moderately]	Malted barley flour is also called malt. Barley grains are allowed to germinate to produce enzymes that convert the grain's starches into sugars. The germinated grains are then ground into flour. If you're old enough to remember malted milk shakes, malt was an important ingredient, as well as in malted milk balls, such as Whoppers. Malted barley flour is often added to bread because it gives the yeast more nutrients (mostly sugar) and provides the bread with a different taste. The malted barley flour may serve to provide enzymes to help condition the dough and it doesn't add any flavor or color. The FDA allows up to 0.75% barley flour in bread flour. [See Barley Malt Extract]
Maltodextrin [Highly]	Maltodextrins are actually a mixture of semi-sweet substances that result from the incomplete industrial breakdown of starch (usually corn starch) by acid or enzymes. These products are used as bulking agents and can impart a mild to moderate sweetness to foods. Maltodextrin is not a pure substance, but is actually a complex mixture of dozens of related substances, so maltodextrin is just a generic term.
Medium Chain Triglycerides (MCT) [Highly]	Medium chain triglycerides (MCTs) are composed of molecules with shorter, i.e., medium chains, of carbons. They are man-made fats. Because of their shorter length, MCTs are more easily digested and absorbed in the GI tract than natural, long-chain triglycerides (fats and oils). They are very popular additives in keto foods, which are

INGREDIENT NAME [PROCESSING INFO]	EXPLANATION
	typically high in fats.
Methylcellulose [Synthetic; Extremely]	This chemical is prepared from wood or cotton by treatment with alcohol. It's used as a binder, thickener, dispersing agent, and emulsifying agent.
Microcrystalline Cellulose [Synthetic; Extremely]	This additive is also called cellulose gel. This ingredient is made by the partial hydrolysis of cellulose (wood fiber). It is used as a filler, texturizer, an anti-caking agent, a fat substitute, an emulsifier, an extender, and a bulking agent in food production.
Mixed Triglycerides [Highly]	[See Triglycerides]
Modified Butter Oil [Extremely]	This ingredient term is not clear but likely refers to butter which has been enzymatically converted into product that provides beneficial properties to a food manufacturer such as liquidity, shelf stabilization, etc.
Modified Cellulose [Highly]	Cellulose extracted from wood pulp or cotton and chemically processed with acids or alkali is added as a creaming agent or thickener to many commercial foods. There are a variety of chemical and physical modifications used but generally specific modifications are not indicated on food labels.

INGREDIENT NAME [PROCESSING INFO]	EXPLANATION
Modified Corn Starch [Extremely]	The word modified implies that the food starch, obtained from a variety of plants, has been physically, chemically or enzymatically altered. This ingredient is one of the most generic items found on food labels. The FDA does not require food companies to reveal the type of modified food starch that is used. It may be made from dozens of different chemicals that have been reacted with the corn starch. Some of these chemicals are very hazardous and the FDA specifies acceptable residue levels for these chemicals. Modified corn starches are used in food products to contribute a wide variety of favorable physical properties, such as stabilizing the starch against temperature changes, serving as a thickening agent or emulsifier, making the starch more dissolvable, and for many other reasons. The modified food starches are usually synthesized in a manufacturing plant using such hazardous chemicals as propylene oxide, acetic anhydride, sulfuric acid, bleach, etc. The FDA approves each modified starch as safe for human consumption; most of them are not natural. The fact that the food manufacturer is not required to reveal the type of modified food starch in the cereal product is not helpful to the consumer who has no idea what they are eating. Here are some examples of modified food starches that are lumped under that generic title: bleached starch, acetylated starch, hydroxypropylated starch, and starch sodium octenyl succinate. That innocent little phrase, "modified corn starch", represents a plethora of industrial modifications, mostly chemical.
Modified Food Starch [Extremely]	[See Modified Corn Starch as an example]
Modified Guar Gum [Synthetic;	Modified forms of guar gum are available commercially, including enzyme-modified, cationic and hydroxypropyl

INGREDIENT NAME [PROCESSING INFO]	EXPLANATION
Extremely]	guar gum. [See Guar Gum]
Modified Potato Starch [Extremely]	Like other food starches, potato starch can be physically and chemically modified in a variety of ways. See Modified Corn Starch.
Modified Tapioca Starch [Extremely]	Like other food starches, tapioca starch can be physically and chemically modified in a variety of ways. See Modified Corn Starch.
Modified Wheat Flour [Extremely]	Like other food starches, wheat starch can be physically and chemically modified in a variety of ways. See Modified Corn Starch.
Mono-Diglycerides [Synthetic; Extremely]	[See Mono- and Diglycerides]
Mono- and Diglycerides [Synthetic; Extremely]	These chemicals are derived from animal and plant fats (triglycerides) and can be made from a variety of sources. They can act as dough conditioners, emulsifiers (agents for mixing water and oil), or stabilizers. This is a very common food additive and often listed in baked goods. It is estimated that every American consumes about half-pound of these additives each year. Since mono- and diglycerides are found in normal fats, they are considered harmless as an additive, but they are still synthetic chemicals manufactured in a factory.
Monocalcium Phosphate [Synthetic; Extremely]	Also known as calcium dihydrogen phosphate, this substance serves as a buffering agent as well as a leavening agent in self-rising flour. If a second acidic component is added, such as sodium bicarbonate, the leavening agent can be described as a double-acting baking powder. One agent acts quickly when the batter is prepared and the second agent kicks in during heating.
Monoglycerides [Synthetic; Extremely]	[See Mono- and Diglycerides]

INGREDIENT NAME [PROCESSING INFO]	EXPLANATION
Monopotassium Phosphate [Synthetic; Extremely]	This chemical, a white powder, is used as a buffer in foods.
Monosodium Glutamate [Synthetic; Extremely]	Usually called MSG, this additive is the main ingredient in the seasoning called Accent and is often found in Chinese food. Its purpose is to boost savory flavor or umami in foods. This food additive has been around for a long time and was one of the original GRAS chemicals grandfathered by the FDA. It's not recommended for pregnant or lactating mothers, infants, or small children. It can be found in other found ingredients that contain glutamate such as hydrolyzed soy products, yeast extracts, whey or soy proteins, whey or soy isolates, and in any ingredient listed as "glutamate." This additive has a controversial history in the food industry, but, to date, despite numerous anecdotal reports over several decades, there are no conclusive medical studies showing that it is a danger to human health.
Natamycin [Highly]	Also called pimaricin, this biological agent and food preservative is a fungicide. It is applied to the surface cuts and slices of cheese to inhibit mold. It's obtained from a type of Strep bacterium. The chemical is also used as a medicine to treat fungal infections around the eye.
Natural Beef Flavor [Highly]	This is a proprietary product that McDonald's uses to add meat-like flavor although there is no beef in it. The main components are hydrolyzed wheat and milk proteins. [See Hydrolyzed Corn, Soy, Wheat.]

INGREDIENT NAME [PROCESSING INFO]	EXPLANATION
Natural Flavor(s) [Moderately]	This is the fourth most common ingredient on food labels. What distinguishes a natural flavor from an artificial flavor? The main difference is that natural flavors come from natural sources, not artificial sources, but the key flavor ingredient(s) may be extracted out of a natural product, then purified, and finally added back to the food. People should not get taken in by that word "natural" since most natural flavors are highly processed. Food companies are not required by the FDA to specify what flavor additives are used, hence the generic phrase "natural flavor." The consumer is left in the dark.
Natural Flavors with Extractives of Paprika [Moderately]	[See Extractives of Paprika]
Natural Smoke(d) Flavor or Flavoring [Highly]	The condensed products from the destructive distillation of wood are called "liquid smoke" or "pyroligneous acid". There are no standards of identity, prescribed production methods, or tests which distinguish between liquid smoke and pyroligneous acid; they can be considered to be the same. However, the numerous variables that are manipulated during pyrolysis do lead to a wide range of compositions of the condensates. In addition, implementation of many further processing steps by concentration, dilution, distillation, extraction, and use of food additives has led to the many hundreds of unique products on the market worldwide.
Natural Spice Extractives [Highly]	[See Spice Extractives]
Neotame [Synthetic; Extremely]	Neotame is a high-intensity, artificial, non-caloric sweetener that is incredibly 8,000 times sweeter than table sugar. It was invented by the same company, G. D. Searle & Company (now part of Pfizer) that introduced aspartame to the marketplace, but it's use in foods is

INGREDIENT NAME [PROCESSING INFO]	EXPLANATION
	more limited. It was approved as an additive in 2002.
Niacin [Synthetic; Extremely]	This is the chemical name for Vitamin B3 and is used to enrich white bread products.
Oleoresin Carrot [Highly]	A concentrated flavoring agent derived from the solvent extraction of carrots using alcohol.
Oleoresin Paprika [Highly]	[See Extractives of Paprika]
Oleoresin Turmeric [Highly]	[See Extractives of Turmeric}
Olive Oil [Highly]	This oil comes in various grades from virgin to refined. Most likely, due to price, pure virgin olive oil is not used in the processed or fast-food industries. The cheaper refined oil is obtained from any grade of the virgin oil by methods that involve the removal of color, odor, and flavor giving a final product which is tasteless, colorless, and odorless plus low in free fatty acids.
Organic Natural Flavors [Moderately]	These flavoring agents come from natural substances that were produced according to the requirements of the USDA Organic Standard rules. See Natural Flavors.
Palm Kernel Oil [Moderately]	Palm kernel oil is a cheaper, by-product of palm oil production and goes through additional refining steps. Residual palm seeds are cracked and deshelled. The kernels then undergo crushing, and the isolated crude oil is further refined.
Palm Oil [Moderately]	Palm oil is more expensive than palm kernel oil, both of which are produced in Asia. It's a replacement for partially hydrogenated fats from corn and soy oils. The food industry used to rely on those fats to keep processed foods from developing rancid odors and flavors, but, based on an FDA ruling, partially hydrogenated fats were phased out in 2018 due to the

INGREDIENT NAME [PROCESSING INFO]	EXPLANATION
	harmful presence of trans fats, which were found to contribute to heart disease. A problem with the use of palm oil is the fact that, in some countries where it is produced, valuable rain forest habitats are being destroyed to propagate the crop. Here are the steps for refining palm oil: (1) sterilization (cooking) of the palm fruits to stop enzyme activity and help release the oil in the next step; (2) steam digestion to release the oil; (3) the oil is separated from pulp using an expeller machine; (4) clarification to remove particles; and (5) storage.
Paprika Extract (Blend) [Moderately]	[See Extractives of Paprika]
Pea Protein [Moderately]	[See Fava Bean Protein]
Pea Starch [Highly]	Pea starch is a by-product of the production of pea protein isolate. The processing steps to produce the final product are fairly complex and proprietary.
Pectin [Highly]	This material is a natural product. It is used as a gelling agent for jams and jellies and as a stabilizer in beverages. Pectin is typically produced from dried citrus peels or apple pulp. The gel is extracted into solution using hot dilute acid which is then concentrated under vacuum. The pectin is precipitated with the addition of alcohol. The dried product is a white to light brown powder.
Phospholipid Enzyme [Highly]	Also called a phospholipase. It is an enzyme produced from a genetically modified organism, possibly the mold Aspergillus niger, in a bioreactor. It has been used to de-gum vegetable oils and fats. In flour making, this additive can improve the stability and holding capacity of dough.
Phosphoric Acid [Synthetic; Extremely]	Phosphoric Acid, also known as orthophosphoric acid, is a weak acid used as an acidifying agent in soft drinks. Medically, it can cause dental erosion and contribute to kidney stone formation. Long-term use has been linked

INGREDIENT NAME [PROCESSING INFO]	EXPLANATION
	to osteoporosis in middle-aged women.
Polyglycerol Esters of Fatty Acids [Synthetic; Extremely]	This additive is a mixture of chemicals formed by the reaction of fatty acids from vegetable oils and fats with polyglycerols. They are used as emulsifiers and stabilizers.
Polysorbate 60 [Synthetic; Extremely]	Found in baked goods, this chemical is a synthetic emulsifier obtained from the chemical combination of polyethoxylated sorbitan with stearic acid. It's also used as a wetting agent, a stabilizer, and dispersing additive. It may be contaminated with 1,4-dioxane, a nasty carcinogen. The sorbate in the name comes from sorbitol, a sugar alcohol.
Polysorbate 80 [Synthetic; Extremely]	It's an amber/golden-colored, viscous liquid made from polyethoxylated sorbitan and oleic acid, a long-chain fatty acid found in animal and vegetable fats. Sorbitans are chemical compounds derived from the dehydration of sorbitol, a sugar alcohol. This additive is used as a defoamer for the fermenting process of some wines, and also as a stabilizer to maintain the creamy texture of ice-cream. It is also used to bulk foods up and keep sauces smooth.
Potassium Benzoate [Synthetic; Extremely]	As an artificial food and beverage preservative, potassium benzoate inhibits mold, yeast, and some bacteria. In soft drinks, it works best at low pH, so it is often combined with acidifying agents.
Potassium Chloride [Synthetic; Extremely]	This natural chemical can be extracted from minerals like potash or from sea water. However, the process for making pure, food grade potassium chloride is complex. This additive can be used as a flavor enhancer, flavoring agent, nutrient supplement, pH control agent, stabilizer or thickener. Sometimes it's used to replace salt (sodium chloride) to lower the sodium content in a food product.
Potassium Citrate	Potassium Citrate regulates acidity in foods and

INGREDIENT NAME [PROCESSING INFO]	EXPLANATION
[Synthetic; Extremely]	beverages as a buffering agent (buffer).
Potassium Iodate [Synthetic; Extremely]	This chemical acts as a dough conditioner and bread improver. A residue of potassium iodate is considered a potential carcinogen.
Potassium Sorbate [Synthetic; Extremely]	This is a common preservative in food. It's a synthetic white, crystalline powder used to inhibit mold, fungi, and yeast. It has a low toxicity.
Potato Dextrin [Highly]	Dextrins are derived from starches, in this case potato starch, which have been broken down into smaller molecules using acid and then dried to give a water-soluble solid.
Potato Flour [Moderately]	Potato Flour is a powder made from ground potatoes. It's commonly used in baking. It can add flavor and texture to baked goods. Whole potatoes are first cooked, usually in large industrial ovens, then dehydrated. Then the dried potatoes are ground into a fine powder. The finished product resembles wheat flour in texture and feel, but it behaves differently in baked goods.
Potato Protein [Highly]	Potatoes have about 2% protein in them. Potato protein is separated from potato flour and extracted to produce an isolate.
Potato Starch [Moderately]	A potato contains starch grains. To extract the starch, the potatoes are crushed and the starch grains get released from the destroyed cells. Then the starch is washed out and dried to produce a powder.
Potato Starch Modified [Extremely]	[See Modified Potato Starch]
Powdered Cellulose [Highly]	[See Cellulose Powder]
Propylene Glycol	This chemical is used in foods to prevent discoloration

INGREDIENT NAME [PROCESSING INFO]	EXPLANATION
[Synthetic; Extremely]	during storage. It can also be used as a defoamer. Large oral doses in animals have been reported to cause central nervous system depression and slight kidney changes.
Propylene Glycol Alginate [Synthetic; Extremely]	This chemical is a white, water-soluble powder that has multiple uses as a stabilizer, thickener, and emulsifier. It's made from the chemical combination of alginic acid, derived from the cell walls of brown algae, and the synthetic chemical propylene glycol. It can be used up to 0.5% by weight in a food product and may cause allergic reactions.
Propylene Glycol Monoesters of Fats and Fatty Acids [Synthetic; Extremely]	These substances are emulsion stabilizers. They are mixtures of propylene glycol mono- and diesters of saturated and unsaturated fatty acids derived from edible oils and fats [There's a considerable amount of chemistry in that last sentence!].
Pyridoxine Hydrochloride [Synthetic; Extremely]	This is the chemical name for vitamin B6 which, by federal mandate, is added to refined flour to fortify it.
Red 40 [Synthetic; Extremely]	This substance is also known as Allura Red. It's an artificial dye that imparts a red color to foods and beverages. It's suspected to cause problems for hyperactive children.
Reduced Iron [Synthetic; Moderately]	Metallic iron in finely divided form, produced by reduction of iron oxide. This is a form of iron sometimes added to foods, such as bread. Also known by its Latin name *ferrum redactum*.
Refined Coconut Oil [Moderately]	Coconut oil is derived from the dried meat of the coconut, which is called copra. The oil from the copra may be cold pressed to produce "virgin" oil. For the refined product, the oil is pressed out using an expeller process, which can heat the oil up. In the final stage, the oil is filtered.

INGREDIENT NAME [PROCESSING INFO]	EXPLANATION
Riboflavin [Synthetic; Extremely]	This is the chemical name for Vitamin B2 and is used to enrich white bread products.
Rice Flour [Moderately]	Rice flour is made by first removing the hull from the rice and then finely grinding it.
Rice Protein [Highly]	Rice Protein powder is made by grinding up rice grains (5 to 7% protein) and treating them with multiple enzymes that separate the starch and fiber from the protein. The protein is then dried and ground into a powder.
Rice Starch [Highly]	Starch is released from rice by first removing protein using an alkaline extraction process. The intact starch is filtered, washed, and dried.
Rosemary Extract [Highly]	This is a specific example of a spice extractive. [See Spice Extractives.]
Sherry Wine Powder [Moderately]	Wine powders are pure wine that is spray dried into a convenient powder form. They are easily rehydrated, with water. This additive is a flavoring agent.
Smoke Flavoring [Highly]	[See Natural Smoke Flavor]
Sodium Acid Sulfate [Synthetic; Extremely]	Also called sodium bisulfate or sodium hydrogen sulfate. As an additive it is used as an acidifying agent and as a leavening agent for cake mixes.
Sodium Acid Pyrophosphate [Synthetic; Extremely]	Used in the baking industry, sodium acid pyrophosphate (SAPP) is a key part of a leavening agent as it combines with sodium bicarbonate (baking soda) to produce carbon dioxide gas to raise the batter up. SAPP is strictly an industrial compound.
Sodium Alginate [Synthetic; Extremely]	Sodium Alginate is chemically derived from a natural product, brown seaweed, that grows in cold water regions. It's known for the property of dissolving in water with strong agitation to give a thick solution, i.e., it

INGREDIENT NAME [PROCESSING INFO]	EXPLANATION
	provides viscosity and emulsification to a mixture.
Sodium Aluminum Phosphate [Synthetic; Extremely]	This chemical is used as a leavening agent for self-rising flours. In baking, it's typically used for the preparation of biscuits, muffins, and sponge cakes. It improves tenderness, moistness, and volume. Often combined with other leavening agents such as monocalcium phosphate and sodium acid pyrophosphate. Note that there are several additives in this glossary that have aluminum in them. That metal is not natural to the human body and has been implicated in brain disorders like Alzheimer's.
Sodium Aluminosilicate [Synthetic; Extremely]	This chemical is an anti-caking additive used in egg and cheese products to prevent clumping.
Sodium Aluminum Sulfate [Synthetic; Extremely]	This chemical is used to bleach flour in the making of white flour.
Sodium Benzoate [Synthetic; Extremely]	This compound is a very common preservative. Although a natural product, food companies use a synthetic form of it. It is active against bacteria and yeast. It is only toxic in large amounts.
Sodium Bicarbonate [Synthetic; Extremely]	[See Baking Powder]
Sodium Caseinate [Synthetic; Extremely]	Sodium Caseinate is chemically made from casein, a milk protein. As an additive in foods, this white powder can serve as an emulsifier, thickener, and stabilizer.
Sodium Citrate [Synthetic; Extremely]	This white synthetic powder, the sodium salt of citric acid, is used as an emulsifier for creams and as a buffer to control acidity.

INGREDIENT NAME [PROCESSING INFO]	EXPLANATION
Sodium Diacetate [Synthetic; Extremely]	Sodium Diacetate is a fungicide and a bactericide used to control mold and bacteria in foods. Sodium diacetate is a mixture of sodium acetate and acetic acid. It is a white, hygroscopic (attracts water), crystalline solid having an odor of acetic acid (commonly known as vinegar).
Sodium Erythorbate [Synthetic; Extremely]	An antioxidant used in pickling brine, meat products, beverages, baked goods, and in cured products, such as bacon, to accelerate color fixing. It limits the formation of nitrosamines in cured meats, which are notoriously carcinogenic.
Sodium Lactate [Synthetic; Extremely]	This chemical acts as a preservative, acidity regulator (buffer), and bulking agent.
Sodium Metabisulfite [Synthetic; Extremely]	This additive serves as a dough conditioner, preservative, inhibits fermentation of sugars, and prevents browning of fruits.
Sodium Nitrite [Synthetic; Extremely]	This chemical has multiple functions in cured meats: preservative, enhances color (the desirable pink hue in meats) and taste, and inhibits rancidity. The International Agency for Research on Cancer has classified this additive as a probable carcinogen in humans.
Sodium Phosphate [Synthetic; Extremely]	This additive comes in 3 forms: monobasic, dibasic, and tribasic or monosodium, disodium, and trisodium. Often the form is not indicated on the food label. It is used as a buffer and effervescent.
Sodium Propionate [Synthetic; Extremely]	Sodium Propionate is used as a preservative in food to inhibit the formation of mold and fungus. Usually found in baked goods, frostings, confections, and gelatin products. It can be found naturally in the body, but it is synthetically made as an additive.
Sodium Saccharin [Synthetic; Extremely]	A synthetic chemical used as an artificial sweetener in foods and beverages since 1879. It is 300 times as sweet as table sugar, but it does leave a bitter aftertaste, which

INGREDIENT NAME [PROCESSING INFO]	EXPLANATION
	can be masked when combined with other sweeteners. The Food & Drug Administration recommends a usage not to exceed 1g per day for a 150-lb person. As a non-nutritive sweetener, it contributes zero calories.
Sodium Silicoaluminate [Synthetic; Extremely]	This compound, a flow agent, is one of a series of amorphous hydrated sodium aluminum silicates. It is present in a product to prevent powdered food from caking, lumping, or aggregating.
Sodium Stearoyl Lactylate [Synthetic; Extremely]	Sodium Stearoyl Lactylate, a cream-colored powder or brittle solid, is also called SSL for short. It makes the gluten in bread stronger and more extensible, meaning the dough is less likely to break or stick during manufacturing. Strengthening the gluten improves the dough's ability to rise, which increases the volume of the finished loaf. It also helps produce softer crumbs and more uniform texture. In the baking industry, it is thought to improve the ability of the bread to "resist abuse." SSL is currently manufactured by the esterification of stearic acid with lactic acid and partially neutralized with either food-grade soda ash (sodium carbonate) or caustic soda (concentrated sodium hydroxide).
Sorbic Acid [Synthetic; Extremely]	This white powder serves as a preservative in foods protecting against mold, fungi, and yeast, especially in cheeses and beverages. It can be naturally isolated from the berries of mountain-ashes, but is more commonly chemically synthesized.
Soy Lecithin [Highly]	Soy Lecithin is one of the 10 most used ingredients in processed foods. Lecithin is a natural product and available from many plant sources. It's found in every living cell in the human body, and is particularly concentrated in the brain, heart, liver, and kidneys. Lecithin can be derived from various plant sources, but soybeans are preferable due to their cheapness. Lecithin,

INGREDIENT NAME [PROCESSING INFO]	EXPLANATION
	in conjunction with xanthan gum, contributes thickness and stability to sauces.
Soy Protein [Highly]	Another name for this food ingredient is soy protein concentrate. It is essentially defatted soy flour without the presence of water-soluble carbohydrates. It's about 70% protein. It is used in a variety of food products such as baked goods, cereals, and as an additive in meats. [See soy protein isolate]
Soy Protein Isolate [Highly]	This ingredient is one of the most refined substances used in the food industry with a protein content of 90% on a dry-solids basis. In a complex process, the isolate is made by stripping soy flour of all other nutrients. It is commonly used as a meat substitute, to increase moisture content, and as an emulsifier.
Soybean Oil [Highly]	This material is a common cooking oil. It also softens bread to make it less elastic. Soybean oil is one of the cheapest and least nutritious oils available. Soybean oil, like most vegetable oils, is extracted from the plant germ using a toxic solvent like petroleum-derived hexane. The resultant solution is clarified and decolorized to produce a water-clear liquid. Then the hexane is stripped off leaving a colorless or lightly-colored oil. Soybean oil is a very processed ingredient.
Spice Extractives [Highly]	This is another generic term. There are two basic spice extractives - essential oils and oleoresins. Oleoresins contain non-volatile materials as well as volatile essential oils. Various kinds of organic solvents are used to prepare oleoresins; the solvents are then removed in accordance with federal regulations once the extraction is completed. The oleoresins are then used as flavoring agents in foods.
Steviol Glycosides [Extremely]	These industrial chemical mixtures are extracted from the natural sweetener found in the stevia plant. The extracts are purified and sold as sweetening agents to food

INGREDIENT NAME [PROCESSING INFO]	EXPLANATION
	companies. These substances are many times sweeter than table sugar (sucrose), so they are only needed in small amounts.
Succinic Acid [Synthetic; Extremely]	This synthetic chemical is prepared from acetic acid and is used as a buffer and neutralizing agent.

INGREDIENT NAME [PROCESSING INFO]	EXPLANATION
Sucralose [Synthetic; Extremely]	An artificial, non-caloric sweetener which is 600 times sweeter than table sugar. Commercial sweeteners containing sucralose are Splenda, Zerocal, Sukrana, and SucroPlus, among others. To replace the 39g of sugar in a 12-oz can of Coke would require only 0.065g or 65mg of pure sucralose ... just a smidgen. Like many of the artificial sweeteners, sucralose was an accidental discovery. The British company Tate & Lyle patented it in 1976. Sucralose got approved in 1998 in the U.S. Sucralose, as an artificial sweetener has several advantages over its competitors. It's non-caloric, so it pretty much gets passed through the body, and it resists breaking down with heat, so it can be used in baking. The process for making sucralose is complex, chemically intensive, and involves very hazardous chemicals. [See text.] It would definitely be a candidate for the most highly refined food ingredient in the history of the modern food industry. Although the starting material is table sugar, there is nothing else natural about the production of sucralose. For many years the marketing firm for sucralose used the slogan "Made from sugar, so it tastes like sugar." The trade group, the Sugar Association, objected to that slogan and won a lawsuit to stop its use. One of the chlorinating agents used in the manufacture of sucralose is phosgene, one of the most poisonous chemicals known to man. In WW1 it was used in chemical warfare. Sucralose, as a sweetener, is sold in very dilute mixtures, so the sucralose you buy in the supermarket has a high filler to sweetener ratio.
Sucrose [Moderately]	Sucrose is the chemical name for table sugar. See Sugar.
Sucrose Acetate Isobutyrate [Synthetic;	This material is not a pure substance but a mixture of esters of sucrose with acetic and isobutyric acids. This synthetic material is used as an emulsifier and helps to

INGREDIENT NAME [PROCESSING INFO]	EXPLANATION
Extremely]	distribute and suspend flavor oils in a beverage.
Sugar [Moderately]	Note that in the food industry there are typically two sources of sugar: cane sugar and beet sugar. Most beet sugar in the United States comes from genetically modified beets, so, if you are at all concerned about consuming food ingredients derived from genetically modified plants, then beet sugar should be avoided unless it comes from organic beets. Food companies generally don't indicate the sugar source. It's a very processed food. Here is a brief description of the process for making cane sugar. The cane, which has 12 to 14% sucrose in it, is cut and taken to a sugar mill. The cane juice is extracted out by being crushed between large rollers. The sugar mixture is then clarified to remove wax, fat, and gums. Then the sugar is concentrated by boiling under vacuum to remove water until sugar crystals form. The raw sugar crystals are spun down in a centrifuge to remove the last amount of water to give a golden raw sugar which is 96 to 98% sucrose. To produce white sugar, the raw material goes to a refinery to remove the last traces of molasses. Then the sugar solution is passed through carbon filters (most likely bone char made from animal bones) to remove the remaining color. The resultant white syrup is vacuum concentrated and the sugar crystallized. The crystals are centrifuged to remove the last traces of water and impurities and then dried.
Sugar Cane Fiber [Highly]	Made from the dry cell wall or fibrous residue remaining after expression or extraction of sugar juice from sugar cane. It consists mostly of cellulose and hemicellulose. The production process consists of several steps, including (1) chipping, (2) alkaline digestion, (3) removal of lignans and other non-cellulosic components, (4) bleaching of purified fibers, (5) acid washing, and (6) final neutralization.

INGREDIENT NAME [PROCESSING INFO]	EXPLANATION
Sulfur Dioxide [Synthetic; Extremely]	This chemical is found in nature as a gas, e.g., product of volcanic eruptions. Most sulfur dioxide is produced by the combustion of elemental sulfur. Sulfur dioxide is sometimes used as a preservative for dried apricots, figs, and other dried fruits, owing to its antimicrobial properties and ability to prevent oxidation. As a preservative, it maintains the colorful appearance of the fruit and prevents rotting. Some people have sensitivities to this chemical.
Sunflower Lecithin [Moderately]	Sunflower lecithin is made by dehydrating sunflowers and separating them into three parts: oil, gum, and solids. The lecithin comes from the gum. It is processed through a cold press system like the one used to make olive oil. This ingredient serves as an emulsifying agent.
Sunflower Oil [Highly]	Sunflower oil is obtained from sunflower seeds through a refining process that involves solvent extraction, de-gumming, neutralization, and bleaching to remove color.
TBHQ [Synthetic; Extremely]	TBHQ stands for tertiary butylhydroquinone. This synthetic powder is a potent anti-oxidant helping to keep foods from developing off-tastes and colors. It combines with oxygen very rapidly and then turns brown. In large amounts TBHQ is harmful. Ingestion of only a gram produces nausea, vomiting, ringing in the ears, and delirium. It is commonly used in foods such as crackers, microwave popcorn, butter and chicken nuggets.
Tapioca Dextrin [Highly]	Dextrins are derived from starches, in this case tapioca starch, which have been acid degraded into smaller molecules and then dried to give a water-soluble solid.
Tapioca Maltodextrin [Highly]	This maltodextrin additive is made from tapioca starch instead of corn or wheat starch. [See Maltodextrin]
Tapioca Starch [Highly]	This natural substance is obtained from cassava roots in an industrial refining process.

INGREDIENT NAME [PROCESSING INFO]	EXPLANATION
Tapioca Starch-Modified [Extremely]	[See Modified Tapioca Starch]
Tapioca Syrup [Highly]	Tapioca starch is refined using acid or enzymes to produce tapioca syrup in a process similar to the manufacture of corn syrup.
Thiamine [Synthetic; Extremely]	This is the chemical name for Vitamin B1 and is used to enrich white bread products.
Thiamine Hydrochloride [Synthetic; Extremely]	This is a form of Vitamin B1 and is used to enrich white bread products.
Thiamine Mononitrate [Synthetic; Extremely]	This is a form of Vitamin B1 and is used to enrich white bread products.
Titanium Dioxide [Synthetic; Extremely]	This industrial chemical is also used to whiten paint, paper, plastic, rubber (think tire white walls in the old days), glass, and ceramics. Titanium dioxide nanoparticles (TiO_2) are widely used food additives. TiO_2 nanoparticles are used to prevent UV light from penetrating food effectively expanding the shelf life. Additionally, it is used as a colorant to make foods appear white by enhancing their opacity. Products with the highest amounts of this additive are sweets and candies. The health hazards of this chemical are controversial.

INGREDIENT NAME [PROCESSING INFO]	EXPLANATION
Tocopherols (Mixed Tocopherols) [Highly]	Tocopherols are the bioactive components of vitamin E which naturally exists in eight different forms (e.g., alpha-, beta-, gamma- delta-tocopherols and tocotrienols). The most biologically active form of the vitamin is called alpha-tocopherol. It is considered the most active natural form because it is the preferred form of vitamin E transported and used by the liver. Synthetic vitamin E is usually derived from petroleum products. Natural vitamin E from plant oils is generally labelled d-alpha tocopherol, d-alpha tocopherol acetate, or d-alpha tocopherol succinate but can sometimes appear as mixed tocopherols (mixed tocopherols, contain not only d-alpha tocopherol but natural mixtures of beta, gamma, and delta tocopherols). As a food ingredient, tocopherols serve as anti-oxidants to protect the flavors of processed food.
Trehalose [Highly]	This substance is a type of sugar which is half as sweet as table sugar (sucrose). It digests more slowly than table sugar. It can be found naturally (10 to 25% dry weight in mushrooms), but industrially it is extracted from corn starch or biosynthesized using microorganisms.
Tricalcium Phosphate [Synthetic; Extremely]	This chemical is used as a dough conditioner and anti-caking agent.
Triglycerides [Highly]	These additives are derived from fats and oils. Chemically, fats and oils are composed of triglyceride molecules.
Turmeric Extract [Moderately]	[See Extractives of Turmeric]
Turmeric Oleoresin [Highly]	[See Oleoresin Turmeric]
Vanilla Extract [Synthetic;	Vanilla Extract can either be made from natural vanilla obtained from vanilla beans (expensive) or it can be

INGREDIENT NAME [PROCESSING INFO]	EXPLANATION
Extremely]	synthesized in a factory (inexpensive). Most likely, fast food restaurants use the artificial variety.
Vegetable Fiber [Moderately]	The cellulosic portion of vegetables, such as carrots, beets, kale, and other plants are examples of vegetable fiber. These materials are readily obtained from the insoluble residues left over from vegetable juicing.
Vegetable Mono & Diglycerides [Synthetic; Extremely]	[See Mono and Diglycerides]
Vegetable Oil [Highly]	Vegetable oil is the generic name for plant-based, extracted oils. Usually, it is represented by a blend of oils such as corn, soy, and canola.
Vinegar [Moderately]	This is a very safe and common food ingredient that has been around for a very long time. It's used for flavoring, acidifying, and preserving. However, the cheap version, white vinegar, is an industrial product. It's a 4 to 6% solution of acetic acid. The latter chemical is derived from the fermentation of alcohol (ethanol) followed by a distillation process. The making of vinegar is a fairly complex, industrial process.
Vinegar Solids [Moderately]	Also known as "vinegar powder" or "dried vinegar" this additive is made by spraying maltodextrin with vinegar and letting it dry. Vinegar powder can be made with any type of vinegar. For example, apple cider vinegar powder as well as malt vinegar powder is manufactured.
Vital Wheat Gluten [Moderately]	[See Wheat Gluten]
Vitamin A [Synthetic; Extremely]	Although Vitamin A is found naturally in many foods, the substance added to industrial foods is factory synthesized.
Vitamin A Palmitate	[See Vitamin A]

INGREDIENT NAME [PROCESSING INFO]	EXPLANATION
[Synthetic; Extremely]	
Vitamin C [Synthetic; Extremely]	Although Vitamin C is found naturally in many foods, the substance added to industrial foods is factory synthesized. [See Ascorbic Acid]
Vitamin D3 [Synthetic; Extremely]	Although Vitamin D3 is found naturally in many foods, the substance added to industrial foods is factory synthesized.
Wheat Flour [Moderately]	If the term "whole wheat" does not appear in the name, then most likely cheap, white flour is being used. Wheat flour is a significantly processed food for the following reasons: (1) The nutritious parts of the wheat, the bran (fiber) and the germ (oil/vitamins) are stripped away; (2) To provide a bright, white color, the wheat flour is bleached with oxidizing agents, some of which are dangerous to health; and (3) The flour is fortified with vitamins and minerals as mandated by the federal government to replace what was lost in the processing of the wheat. In that case, the product is known as enriched wheat flour.
Wheat Gluten [Moderately]	Gluten is that portion of the wheat flour that's left after the starchy component is removed. Wheat gluten is very high in protein. It gives structure to baked goods. Gluten is the sticky, stretchy material that forms when wheat flour and water are mixed. Gluten increases the dough's ability to rise. It also increases the bread's structural stability and chewiness. Although gluten is naturally present in flour, it requires substantial kneading to be released. Adding wheat gluten is a shortcut. It's considered a manufacturing or processing aid. A small percentage of the population has a sensitivity to wheat gluten, and for people with celiac disease, consuming gluten can be life threatening.

INGREDIENT NAME [PROCESSING INFO]	EXPLANATION
Wheat Protein Isolate [Moderately]	Wheat Protein Isolate is prepared by removing starch from wheat flour, then carefully drying the remaining high protein fraction to retain the viscoelastic properties. This additive is highly effective in applications where strength, elasticity, exceptional functionality and high protein content are all required. Wheat protein is added to bread dough to obtain a better rise/strength/chewiness.
Wheat Starch [Highly]	Typically, wheat starch is obtained from wheat flour by mixing flour with water and separating the starch slurry by some form of centrifuging. The precipitated starch is then washed and dried.
Whey [Moderately]	During the cheese-making process, the liquid remaining after the cheese is coagulated is whey. It has various uses such as a texturizer, processing aid, and nutritional extender.
Whey Powder [Moderately]	During the cheese-making process, the liquid remaining after the cheese is coagulated is whey. When water is removed from this product, then whey powder is produced. It has various uses such as a texturizer, processing aid, and nutritional extender (particularly in protein supplements).
Whey Protein Concentrate [Moderately]	After milk is curdled to make cheese, the watery portion of the milk is whey. In the old days, there wasn't much use for whey, and it was routinely disposed of as a waste product. But in the 1980s new technologies, such as ultrafiltration, were developed that allowed milk producers to separate the small amount of soluble, globular protein in the whey from the water. Whey protein concentrate is used as a water binder.
Whey Solids [Moderately]	[See Whey Powder]
White Vinegar	[See Vinegar]

INGREDIENT NAME [PROCESSING INFO]	EXPLANATION
[Moderately]	
White Wine Vinegar [Moderately]	[See Vinegar]
Xanthan Gum [Synthetic; Extremely]	This is a plant-based thickener added to food products to provide some viscosity (thickness). Most commonly, the xanthan gum is derived from the fermentation of corn-derived dextrose, and there's a complex industrial process to make the final product, which is a dry, white powder.
Yeast Extract [Highly]	See Autolyzed Yeast Extract.
Yellow 5 [Synthetic; Extremely]	This artificial dye is also known as tartrazine. It is manufactured in a chemical plant. Aspirin-sensitive people may react to the consumption of this dye.
Yellow 6 [Synthetic; Extremely]	This artificial dye is also known as Sunset Yellow FCF. It is manufactured in a chemical plant. Some people may have allergic reactions to it.

ABOUT THE AUTHOR

Mel Weinstein is a retired chemist currently living in Central Illinois. During the course of his professional career, he has taught chemistry at the community college level, spent 20 years as an analytical chemist in a research center for a multinational food ingredients manufacturer (Tate & Lyle), studied food science, and pursued a lifelong interest in examining ingredients in industrial foods and the nutritional impacts on human health.

In 2016 he started a podcast called "Food Labels Revealed" to share knowledge of the processed food industry and to highlight the ever-growing trend of artificial foods in the American food supply. The podcast can be found at the following internet location and is available through major podcast players like Apple, Stitcher, and Pandora.

https://foodlabelsrevealed.podbean.com/?source=pb

He also has a Facebook page called "Food Labels Revealed Podcast" where he posts commentaries and links to on-line articles about the processed food industry.

https://www.facebook.com/prophetofprocessedfood

The website for "Food Labels Revealed" can be found at:

www.foodlabelsrevealed.com

He can be reached at:

FoodLabelsRevealed@gmail.com

ACKNOWLEDGEMENTS

Generally, I was greatly encouraged to write this book from the words of praise and positive reviews from the listeners of the Food Labels Revealed Podcast over the last 6 years. From their feedback, I learned that there was an audience for this type of information.

Also, I am grateful to the many authors who have chosen to research and write about the health issues associated with ultra-processed foods, junk foods, and fast foods. They have greatly inspired me to contribute to that subject. Among the authors who have fueled my interest include, but are certainly not limited to: Joanna Blythman, Randall Fitzgerald, Joel Fuhrman, Michael Moss, Marion Nestle, Robyn O'Brien, Mark Schatzker, Michele Simon, Melanie Warner, Bee Wilson, and Ruth Winter.

Specifically, as regards the contents of the book, I want to thank the following people, whose uncompensated efforts were essential in completing this work.

Joyce Throneburg, long-time friend, proofed and edited most of the book. Her ability to catch errors, provide feedback on syntax and my quirky writing style, and suggestions for content were greatly appreciated. She took the time out of her busy personal life to review the book which went above and beyond a simple favor for a friend. She even dusted off her dormant math skills to check the calculations in the book.

Mike Grace took on the mind-numbing task of checking the entries to the Glossary. As a fellow chemist with excellent analytical skills, he was a good choice for the job.

My wife, **Sue Weinstein**, a former English educator with a masters in communication, edited the final chapters of the book. Thanks dear for your contribution to and support of this project.

My gratitude to my friend **Terry Ankrom** who helped me check the viability of the hyperlinks. His attention to detail was much appreciated.

INDEX

A

Acceptable Daily Intake (ADI) · 136
acetic acid · 114, 189, 310, 330, 333, 339
acetic anhydride · 75, 319
acidifying agent · 149, 325, 326, 329
acrylamide · 203
activated carbon · 123
ad libitum · 215
Additives · 21, 49, 294, 297
aeration · 76, 312
American Chemical Society · 196
ammonia · 123
annatto · 56, 58, 59, 145, 152, 159, 265, 267, 298, 305
anti-caking agent · 107, 190, 303, 305, 318, 338
anti-oxidant · 106, 107, 108, 121, 135, 136, 180, 189, 299, 302, 336, 338
BHA · 65, 106, 107
BHT · 65, 106, 107
butylated hydroxyanisole · 65, 106
butylated hydroxytoluene · 65, 107
artificial color · 41, 79, 80, 81, 120, 312, 328
Blue 1 · 77, 79, 80, 82, 164, 301, 312
Brilliant Blue · 80, 301
Red 40 · 40, 78, 79, 80, 82, 164, 312, 328
tartrazine · 80, 342
Yellow 5 · 40, 78, 79, 80, 82, 342
artificial flavor · 67, 77, 79, 80, 84, 164, 166, 299, 312, 322
artificial substances banned · 21
cyclamate · 21
Ethyl Acrylate · 21
Partially hydrogenated fats and oils · 21
Violet 1 · 21
Yellow 1 · 81
Yellow 2 · 81

ascorbic acid · 37, 43, 65, 101, 107, 134, 136, 299, 302, 310
Vitamin C · 65, 101, 107, 299, 340
ascorbyl palmitate · 45, 104, 106, 107, 108, 299
palm oil · 107, 323, 324
palmitic acid · 107
autolyzed yeast extract · 56, 58, 59, 93, 110, 118, 299, 342
azodicarbonamide · v, 93, 95, 96, 97, 101, 300

B

bactericide · 61, 114, 134, 330
benzene · 134
beta carotene · 70, 118, 120, 121, 132, 301
carotenoids · 120
Vitamin A · 57, 119, 120, 132, 145, 153, 160, 224, 340
Beyond Meat Italian sausage · 128
Big Breakfast® with Hotcakes · 55, 69, 71, 72, 271
binder · 107, 128, 147, 305, 318, 342
black olive · 113
bleach · 75, 95, 319, 330
ascorbic acid (Vitamin C) · 37, 43, 65, 101, 107, 134, 136, 299, 302, 310
benzoyl peroxide · 101
bleaching agents · 101
calcium peroxide · 96, 97, 101
chlorine · 101, 121
chlorine dioxide · 101
botulism bacteria · 115
bread
enrichment · 84, 98, 99, 102, 113, 140
fortification · 98, 102, 312, 316
brine · 113, 330
bulking agent · 107, 190, 305, 318, 330
butylated hydroxyanisole (BHA) · 65

butylated hydroxytoluene (BHT) · 65

C

calcium disodium EDTA · 93, 118, 132, 133, 134, 136, 145, 152, 302, 303
calcium peroxide · 93, 95, 96, 97, 101, 302
calcium propionate · 93, 152, 159, 161, 265, 267, 302
Canadian Food Expenditure Survey (FOODEX) · 211
cardiovascular diseases · 19, 185, 204, 256
carnauba wax · 164, 166, 304
  Brazil Wax · 166
  carnauba palm · 166, 304
  confectioner · 167
  glaze · 167
  lac beetle · 167
  Palm Wax · 166
  sticklac · 167, 168
carrageenan · 45, 70, 77, 104, 145, 147, 148, 152, 156, 159, 205, 304
categories of processing · 34
  extremely processed · 25, 32, 33, 34, 37, 38, 41, 44, 66, 78, 83, 85, 86, 99, 105, 133, 146, 154, 191, 266, 268, 269, 297
  highly processed · 233
  highly processed · 18, 19, 38, 41, 42, 44, 45, 47, 49, 52, 54, 58, 112, 120, 173, 183, 186, 191, 201, 217, 218, 221, 222
  highly processed · 255
  highly processed · 260
  highly processed · 261
  highly processed · 262
  highly processed · 263
  highly processed · 268
  highly processed · 322
  lightly processed · 37, 38, 41, 45, 47, 52, 95, 146, 191, 268, 297
  moderately processed · 38, 40, 41, 191, 268

unprocessed · 25, 34, 37, 38, 39, 41, 44, 45, 47, 146, 187, 188, 191, 203, 205, 211, 215, 216, 217, 266, 268, 269
categories of research
  experimental · 192, 195, 198
  observational · 192, 194, 195, 197, 199, 201, 202, 212, 214, 219
cellulose · 45, 57, 70, 84, 94, 104, 106, 107, 126, 145, 152, 159, 265, 267, 304, 305, 318, 327
Center for Science in the Public Interest · 135
Centers for Disease Control (CDC) · 22, 177
  National Center for Health Statistics · 22
Cheetos® · 40
chelating agent · 134
  sequestering · 134
Chemistry for the Consumer · 4
chloromethane · 128
Cinnabon® Stick · 91, 131, 132, 133, 135, 136, 281
citric acid · 40, 56, 62, 70, 84, 93, 110, 118, 126, 132, 145, 147, 149, 152, 155, 159, 164, 305, 330
  aspergillus niger (A. niger) · 149
  n-octyl alcohol · 149
  tridodecyl amine · 149
coating · 107, 128, 167, 304, 305, 306
colorectal cancer · 115, 200, 201
coloring agent · 21, 74, 80, 301, 304
conditioning agent · 96
conflict of interests (COIs) · 229
corn · 2, 5, 7, 9, 11, 40, 66, 73, 75, 76, 82, 83, 87, 121, 150, 168, 183, 189, 190, 217, 221, 223, 225, 229, 235, 242, 245, 253, 306, 307, 308, 315, 318, 319, 324, 337, 338, 339, 342
corn syrup · 2, 5, 7, 73, 82, 83, 84, 121, 150, 183, 190, 221, 229, 307, 337
corn syrup solids · 56, 110, 118, 120, 121, 145, 153, 159, 307
COVID-19 · 69, 180
  pandemic · 180
  polysorbate 80 (Tween 80) · 69
  vaccine · 69

Crispy Chicken Caesar Salad · 91, 94, 95, 288

Crystalline dextrose · 73

curing process of meat · 114, 115, 121, 149, 156, 189, 305

accelerator · 156, 197

D

DATEM · v, 34, 36, 45, 62, 64, 65, 66, 104, 110, 118, 132, 307, 308

Diacetyl Tartaric Acid Esters of Mono- and Diglycerides · 65, 307

death statistics · 177

degree of industrialization · 34, 41, 44, 133, 191, 261, 262

Degree of Processing (DP) · 33, 269

dextrose · 45, 56, 62, 70, 72, 77, 104, 110, 118, 145, 153, 159, 164, 308

Dietary Goals for the United States · 6

Dietary Guidelines for Americans · 6, 225, 226, 256, 263

dimethyl formamide · 122, 123

dimethylpolysiloxane · v, 93, 95, 97, 309

dimethylpolysiloxane (PDMS)

Antifoam A · 97

dimethicone · 97

disability-free life expectancy index · 182

disodium guanylate & disodium inosinate · 58

dispersant · 128

dough conditioner · 65, 66, 76, 95, 155, 298, 300, 302, 303, 307, 316, 320, 326, 331, 338

E

EDTA · 93, 118, 132, 133, 134, 136, 145, 152, 294, 302, 303

ethylene diamine tetraacetic acid · 133

emulsifier · 65, 68, 75, 76, 107, 128, 131, 147, 149, 150, 156, 184, 190, 205, 295, 301, 305, 307, 308, 309, 311, 315, 318, 319, 320, 325, 327, 330, 333, 335, 336

Environmental Protection Agency (EPA) · 134

Environmental Working Group (EWG) · 141, 232, 240, 254, 257

enzymes · 34, 36, 40, 45, 56, 63, 83, 93, 104, 110, 118, 126, 132, 145, 147, 150, 153, 159, 163, 265, 267, 295, 299, 300, 308, 310, 316, 317, 318, 320, 324, 325, 328, 337

amylases · 150

rennet · 150

essential oils · 73, 333

ethyl acetate · 122

European Food Safety Authority (EFSA) · 135

experimental designs

case study · 194

cohort · 194, 199, 201, 206, 207, 208, 209, 212, 213, 219, 220, 256, 263

cohort studies · 194, 219, 220

confounding errors · 195

cross-over RCT · 193

cross-over study · 215

epidemiological studies · 185, 195

experimental errors · 195

information bias · 195

observational designs · 194, 195

prospective · 194, 195, 196, 199, 219, 220, 230, 256, 263

random controlled trial (RCT) · 192, 214

random errors · 195

random variation · 195

retrospective · 194, 195

selection bias · 195

systematic error · 195, 202

extender · 107, 305, 318, 341

extractives of black pepper · 70, 72, 73, 311

F

Fair Packaging and Labeling Act · 11
fast food · 2, v, 3, 4, 10, 17, 18, 19, 21,
  22, 23, 25, 26, 27, 28, 29, 31, 43,
  49, 51, 86, 91, 96, 99, 106, 112,
  120, 127, 140, 146, 176, 191, 194,
  208, 221, 230, 239, 255, 260, 261,
  262, 263, 264, 339
fast food restaurants · 17, 28, 43, 86,
  91, 112, 120, 127, 261, 264, 339
fat substitute · 7, 8, 107, 305, 318
Federal Trade Commission (FTC) · 123
fermentation process · 60, 121, 135,
  149
fiber · 5, 12, 83, 101, 107, 108, 147,
  155, 180, 202, 205, 207, 211, 215,
  231, 237, 244, 305, 306, 316, 318,
  328, 339, 340
  cellulose · 107, 108, 128, 156, 304,
    305, 318, 335
  hemicellulose · 108, 335
  insoluble fiber · 107, 215, 305
  pectin · 108, 148, 295, 324
  psyllium husk · 108
filler · 107, 305, 318, 334
Firmenich · 80, 299
flavor agent · 59, 74
flavor enhancers · 59
flavorists · 79, 80
folic acid (Vitamin B9) · 98
Food & Drug Administration (FDA) ·
  7, 20, 30, 260, 299
  FDA · 8, 17, 20, 30, 33, 51, 58, 61,
    62, 65, 66, 67, 75, 76, 79, 80,
    81, 96, 97, 108, 115, 122, 124,
    127, 129, 133, 134, 135, 136,
    148, 150, 163, 168, 173, 174,
    231, 232, 260, 261, 262, 299,
    300, 307, 317, 319, 321, 322,
    324
food addiction · 29, 216, 218
food additives · 6, 9, 15, 17, 20, 25, 49,
  51, 60, 61, 122, 148, 185, 234, 252,
  260, 261, 262, 322, 337
Food Labels Revealed · 15, 227, 264,
  343

Food Labels Revealed Podcast · 227,
  343
food processing · 25, 39, 186, 188, 228,
  255, 257
formaldehyde · 97
freeze-thaw stability · 76
Frito-Lay · 8, 252
  Wow chips · 8
front-of-package food scoring systems ·
  v
front-of-package labels · 254, 263
  Nutri-Score · 237, 238, 239, 240,
    242, 250, 252, 253, 257, 263
fructose · 2, 5, 7, 73, 82, 83, 162, 183,
  190, 221, 313, 316, 317
fumaric acid · 153, 154, 155, 159, 265,
  267, 313
fungicide · 61, 114, 134, 321, 330

G

gelatin · 74, 84, 86, 128, 148, 313, 315,
  316, 331
Generally Recognized as Safe (GRAS) ·
  20, 51, 261
  GRAS · 20, 51, 96, 127, 129, 163,
    261, 321
genetically modified organisms (GMOs)
  · 130
George McGovern · 6, 224
  U.S. Senate Select Committee on
    Nutrition and Human Needs · 6
George Zaidan
  Zaidan · 196, 197
Givaudan · 80
glucose · 66, 72, 82, 83, 113, 149, 162,
  163, 183, 192, 203, 217, 219, 308,
  315, 316
glucose isomerase · 83
glycemic index (GI) · 217
glycemic load (GL) · 217
glycemic responses · 203
glycerol ester of rosin · 164, 166, 168,
  314
  abietic acid · 168
  Ester Gum · 168, 310

Grande Nachos – Steak · 291
GRAS · 20, 51, 96, 127, 129, 163, 261, 321
green olive · 113

H

Healthy Life Years (HLY) indicator · 182
hexanes · 74
high fructose corn syrup
  HFCS 42 · 83
  HFCS 55 · 84
high fructose corn syrup (HFCS) · 5, 82
  high fructose corn syrup · 5, 7, 73, 82, 84, 190
high-intensity sweetener (HIS) · 121
  1,1,2-trichloroethane · 123
  acesulfame potassium · 121
  activated carbon · 123
  ammonia · 123
  aspartame · 121, 297, 323
  dimethyl formamide · 122, 123
  ethyl acetate · 122
  Federal Trade Commission · 123
  McNeil Nutritionals · 122, 124
  methanol · 122, 123
  saccharin · 121
  sodium methoxide · 123
  sucralose · 121, 122, 123, 124, 334
  sucrose-6-acetate · 122, 123
  Sugar Association · 123, 334
  Tate & Lyle Company · 121
  tertiary butylamine · 122
  thionyl chloride · 123
  trimethyl orthoacetate · 122
humectant · 72, 314
hydrogenated soybean oil · 75
hydrogenation · 74
hygroscopic · 114, 330
hyper-palatable formulations · 184

I

iron · 98, 113, 129, 134, 231, 312, 316, 328
  ferric · 113, 328
  ferrous · 113
  Ferrous Gluconate (Iron (II) Gluconate) · 113
  ferrum · 113, 328
  gluconic acid · 113
  reduced iron · 113

K

Keys, Ansel
  starvation study · 193
  The Biology of Human Starvation · 194

L

lactose intolerance · 4, 11
leavening agent · 156, 297, 298, 300, 308, 313, 320, 329
locust bean gum · 70, 78, 145, 153, 154, 155, 156, 159, 317
  carob · 11, 155, 317
  carob tree · 155, 317
  locust tree · 155
longevity · 14, 180
lubricant · 97, 128, 309

M

malic Acid · 57, 58, 60, 126, 317
maltodextrin · 40, 57, 63, 64, 66, 93, 110, 118, 126, 145, 153, 159, 306, 318, 337
McCafé® Mocha Frappé, Medium · 55, 77
McDonald's
  Big Breakfast® with Hotcakes · 55, 69, 71, 72, 271

Big Mac®, Fries, and Chocolate
  Shake · 87
McCafé® Mocha Frappé,
  Medium · 55, 77
Quarter Pounder® with Cheese
  Bacon · 55, 61, 64, 275
Southwest Buttermilk Crispy
  Chicken Salad · 55, 58, 262,
  275
Meat Lover's® Original Stuffed
Crust® Slice · 91, 116, 120
medium-chain triglycerides (MCTs) · 9
meta-analysis · 115, 219
metabolic ward · 215
methanol · 122, 123
methylcellulose · 128, 315, 318
microbiome · 69, 180
microcrystalline cellulose (MCC) · 107
  cellulose powder · 107
modified food starch · 45, 63, 70, 72,
  75, 76, 78, 84, 93, 104, 110, 118,
  126, 153, 159, 313, 319
  acetylated starch · 76, 319
  bleached starch · 76, 319
  hydroxypropylated starch · 76, 319
  sodium octenyl succinate starch · 76
mold inhibitor · 114, 161
Moli-sani Study · 209
mono- and diglycerides · 65, 72, 76, 77,
  307, 309, 311, 320, 321
morbidity · 177, 182
mortality · 177, 179, 182, 206, 207,
  208, 210, 213, 216, 220, 221, 256,
  263

N

Nabisco
  Snackwell cookie cakes · 7
natamycin · 57, 58, 61, 321
National Institute of Health (NIH) ·
  135, 215
National Institute of Health (NIH)
  Clinical Center · 215
  cross-over study · 215
natural color · 81, 82, 120, 121
natural flavor(s) · 67, 322

Nestle, Marion · 227
  Food Politics · 227
NOVA · 187, 188, 190, 191, 200, 202,
  208, 209, 211, 213, 215, 255
  Group 1 Foods · 188
  Group 2 Foods · 189
  Group 3 Foods · 189
  Group 4 Foods · 189
NutriNet-Santé Study · 199, 204, 206,
  256
  acrylamide · 203
  cardiovascular diseases · 19, 185,
  204, 256
  dietary surveys · 199
  glycemic responses · 203
  mortality · 177, 179, 182, 206, 207,
  208, 210, 213, 216, 220, 221,
  256, 263
  risk of developing cancer · 199
Nutrition Facts panel · 12, 16, 17
Nutrition Labeling and Education Act ·
  12

O

obesity · 4, 6, 12, 17, 19, 23, 176, 177,
  185, 186, 190, 197, 203, 211, 212,
  216, 221, 227, 230, 237, 244, 245,
  254, 255, 256, 258, 263, 264, 315
oleoresins · 73, 333
osmotic pressure · 73
oxidizers · 101

P

partially hydrogenated fats · 75, 324
partially hydrogenated fats and oils · 21
Partially hydrogenated fats and oils
  trans fats · 21, 75, 315, 324
pastry flour · 157
pea protein · 63, 126, 128, 129, 324
Pizza Hut · v, 3, 31, 32, 33, 43, 44, 45,
  48, 90, 91, 95, 103, 105, 106, 109,
  111, 112, 115, 120, 125, 127, 128,

131, 133, 137, 143, 221, 261, 277, 297

Cinnabon® Stick · 91, 131, 132, 133, 135, 136, 281

Crispy Chicken Caesar Salad · 91, 94, 95, 288

Meat Lover's® Original Stuffed Crust® (2 slices), Zesty Italian Salad, and Cinnabon Sticks (5 pieces) · 137

Meat Lover's® Original Stuffed Crust® Slice · 91, 116, 120

Super Supreme Personal Pan Pizza® Slice · 91, 109, 112

The Great Beyond Original Pan Pizza® Slice · 91, 125, 128

Tuscani® Creamy Chicken Alfredo Pasta · 91, 103, 106

polysorbate 80 (Tween 80) · 35, 36, 63, 65, 68, 69, 126, 325

potassium nitrate (saltpeter) · 115

preservative · 43, 61, 65, 68, 72, 97, 114, 130, 134, 135, 149, 155, 161, 205, 294, 298, 299, 302, 307, 309, 310, 321, 326, 330, 331, 332, 336

Processed Food Index (PFI) · 34, 37, 38, 41, 42, 43, 44, 52, 54, 57, 58, 63, 71, 72, 78, 85, 94, 105, 111, 119, 127, 132, 146, 154, 160, 165, 173, 191, 261, 262, 265, 269, 270, 277, 289

degree of industrialization · 34, 41, 44, 133, 191, 261, 262

processed meats · 114, 115, 201, 203, 208, 210

curing agent · 115

curing process of meat · 114

Procter & Gamble · 7

fat-free Pringles · 8

Olean · 8

Olestra · 7

sucrose polyester · 7

proprietary formulas · 67, 79, 174, 262

propylene glycol alginate (PGA) · 147, 150

alginic acid · 150, 161, 327

brown algae · 150, 327

propylene glycol · 150, 327

propylene oxide · 75, 319

Q

Quality-Adjusted Life Year (QALY) · 181

Quarter Pounder® with Cheese Bacon · 55, 61, 64, 275

R

rancidity of fat · 99

revolving door · 228, 229, 256

riboflavin · 98

S

Safety Rankings for Food Additives · 21

scoring systems for processed foods · 222

Armour Chicken Bologna · 233

Food Compass · 234, 257

Friedman School of Nutrition Science and Policy · 234, 257

Fritos the Original Corn Chips · 232

Front-of-Pack (FOP) labeling · 240

Health Star Rating (HSR) · 243

Healthy Living · 232

Mexican Daily Dietary Guidelines · 245

Mike Rayner · 237

Nutri-Score System · 237

Open Food Facts · 239, 252, 253, 254, 257

Stéphane Gigandet · 252

Traffic Light Label · 240

Tufts University · 234, 257

semicarbazide · 96, 300

Sensient · 80, 299

shelf life · 13, 61, 67, 74, 114, 134, 161, 173, 262, 296, 303, 310, 311, 337

sodium · 60, 76, 114, 115, 123, 129, 130, 134, 136, 156, 161, 162, 173,

184, 201, 202, 203, 205, 206, 211, 215, 228, 244, 245, 248, 249, 250, 300, 304, 319, 320, 326, 329, 330, 331, 332
sodium acetate · 114, 330
sodium acid pyrophosphate (SAPP) · 154
   baking soda · 156, 300, 329
   carbon dioxide · 156, 300, 329
   sodium bicarbonate · 156, 300, 320, 329
sodium alginate · 161, 162
   alginic acid · 150, 161, 327
   brown seaweed · 161, 329
   calcium alginate · 162
   calcium chloride · 161, 296
   kelp · 161
   sodium carbonate · 162, 332
sodium benzoate and potassium sorbate · 133, 134
sodium bicarbonate · 156, 300, 320, 329
sodium metabisulfite · 126, 128, 129, 331
sodium methoxide · 123
sodium nitrite · 114, 115, 203
sodium propionate · 161
sodium stearoyl lactylate (SSL) · 97
soluble fiber · 155
sorbic acid · 135
Southwest Buttermilk Crispy Chicken Salad · 55, 58, 262, 275
soy protein · 129, 321, 332
specialty beverages · 79, 142
stabilizers · 76, 148, 190, 320, 325, 327
stomach cancer · 115, 206
Substances Added to Food · 20
sucralose · 121, 122, 123, 124, 334
   activated carbon · 123
   ammonia · 123
   dimethyl formamide · 122, 123
   ethyl acetate · 122
   Federal Trade Commission (FTC) · 123
   methanol · 122, 123
   sodium methoxide · 123
   sucrose-6-acetate · 122, 123
   Sugar Association · 123, 334

   tertiary butylamine · 122
   thionyl chloride · 123
   trimethyl orthoacetate · 122
sucrose-6-acetate · 122, 123
sugar · 7, 34, 36, 40, 46, 57, 63, 71, 72, 78, 82, 83, 84, 85, 94, 104, 106, 108, 110, 118, 121, 123, 126, 132, 145, 152, 159, 163, 231, 265, 267, 275, 297, 299, 301, 310, 311, 313, 315, 316, 317, 323, 331, 333, 334, 335, 338
   cane sugar · 11, 82, 301, 311, 335
   Cuba · 82, 251
Sugar Association · 123, 334
sugar cane fiber · 46, 104, 106, 108, 110, 118, 126, 335
sulfite · 114, 129, 130, 205
sulfiting agent · 129, 130
   sodium and potassium bisulfite · 129
   sodium and potassium metabisulfite · 129
   sodium sulfite · 129
   sulfur dioxide · 129, 336
sulfuric acid · 75, 319
SUN Prospective Cohort Study · 207
sunflower lecithin · 126, 128, 130, 336
   acetylcholine · 130
   choline · 130
   phospholipids · 130
**Super Supreme Personal Pan Pizza®
Slice** · 91, 109, 112

T

Taco Bell · v, 3, 31, 32, 33, 43, 44, 142, 143, 147, 154, 161, 165, 166, 169, 221, 261, 265, 289, 297
   Beef Burrito · 169, 289
   Beef Burrito, Grilled Chicken Grande Nachos, and Wild Strawberry Freeze · 169
   Grande Nachos – Steak · 291
   Power Menu Bowl · 143, 146, 147, 150, 292
   Wild Strawberry Freeze · 143, 164, 166, 167, 168, 169, 172, 293

TBHQ · v, 94, 118, 126, 132, 133, 135, 136, 262, 336
   tertiary butylhydroquinone · 135, 336
tertiary butylamine · 122
texturizer · 107, 163, 297, 305, 309, 318, 341
The Great Beyond Original Pan Pizza® Slice · 91, 125, 128
thickening agent · 75, 76, 148, 155, 205, 314, 317, 319
thionyl chloride · 123
trehalose · 159, 162, 163, 338
   Larinus maculates beetle · 162
   manna · 162
   resurrection plant · 163
   trehala manna · 162
   trehalase · 163
triglyceride · 76, 338
trimethyl orthoacetate · 122
Tuscani® Creamy Chicken Alfredo Pasta · 91, 103, 106
Twinkie · 13
   Steve Ettlinger · 13
   Twinkie, Deconstructed · 13

U

ultra-processed foods (UPFs) · v, 15, 140, 176, 182, 183, 184, 185, 186, 189, 192, 196, 197, 201, 221, 227, 228, 230, 249, 253, 255, 256, 257, 260, 263
United States Department of Agriculture (USDA) · 6, 22, 222
   A Guide to Good Eating · 224
   Agricultural Act of 2014 (Farm Bill) · 226
   Basic Seven food groups · 224
   Dietary Goals · 6, 7, 224, 225
   Dietary Guidelines Advisory Committee (DGAC) · 226
   Dietary Guidelines for Americans (DGA) · 225
   Economic Research Service · 22
   Finding Your Way to a Healthier You · 225
   Food for Fitness, a Daily Food Guide · 224
   Food for Young Children · 223
   Food Guide Pyramid · 225
   Food Wheel
      A Pattern for Daily Food Choices · 225
   Hassle-Free Daily Food Guide · 225
   Health and Human Services (HHS) · 226
   How to Select Food · 223
   MyPlate · 225
   MyPyramid Food Guidance System · 225
   Nutrition and Your Health Dietary Guidelines for Americans · 225
   Recommended Dietary Allowances (RDAs) · 224
   Senator George McGovern's Select Committee on Nutrition and Human Needs · 224
USA Framington Offspring Study · 212

V

viscosity · 87, 315, 329, 342
vital wheat gluten (VWG) · 154, 157
   seitan · 157
Vitamin A · 57, 119, 120, 132, 145, 153, 160, 224, 340
vitamin B1 · 98, 101, 131, 337
   beriberi · 98, 101
   Thiamine Monochloride · 98
Vitamin B12 · 131
   cyanocobalamin · 131
vitamin B2 · 98, 328
   riboflavin · 98
Vitamin B2 · 98, 328
vitamin B3 · 101
   Pellagra · 101
vitamin B9 · 98, 313
vitamin C · 65, 101, 107, 299, 340

W

Washington Post · 23
wheat · 66, 68, 75, 76, 96, 97, 98, 99,
  100, 102, 113, 140, 157, 217, 223,
  301, 307, 310, 313, 315, 316, 320,
  322, 326, 337, 340, 341
  alum · 100
  aluminum sulfate · 100, 298
  Assize of Bread and Ale · 100
  Bee Wilson · 100
  bran · 99, 306, 340
  endosperm · 99, 306
  germ · 83, 99, 101, 306, 308, 333,
    340
  Simnal bread · 100
  The Dark History of Food Fraud ·
    100
  wastel bread · 100
  wheat berry · 99

whole wheat bread · 100, 102, 217
whole-grain flour · 100
white bread · 99, 100, 101, 217, 232,
  313, 323, 328, 337
whitening agent · 95, 300
Wild Strawberry Freeze · 143, 164,
  166, 167, 168, 169, 172, 293
World Health Organization (WHO) ·
  96, 115

X

xanthan gum · 87

Y

Yale Food Addiction Scale (YFAS) ·
  217

Made in the USA
Monee, IL
05 January 2023

24528798R00203